HALLELUJAH MOMENTS

Hallelujah Moments

TALES OF DRUG DISCOVERY

Eugene H. Cordes

OXFORD
UNIVERSITY PRESS

OXFORD
UNIVERSITY PRESS

Oxford University Press is a department of the University of Oxford.
It furthers the University's objective of excellence in research,
scholarship, and education by publishing worldwide.

Oxford   New York
Auckland   Cape Town   Dar es Salaam   Hong Kong   Karachi
Kuala Lumpur   Madrid   Melbourne   Mexico City   Nairobi
New Delhi   Shanghai   Taipei   Toronto

With offices in
Argentina   Austria   Brazil   Chile   Czech Republic   France   Greece
Guatemala   Hungary   Italy   Japan   Poland   Portugal   Singapore
South Korea   Switzerland   Thailand   Turkey   Ukraine   Vietnam

Oxford is a registered trade mark of Oxford University Press
in the UK and certain other countries.

Published in the United States of America by
Oxford University Press
198 Madison Avenue, New York, NY 10016

Library of Congress Cataloging-in-Publication Data
Cordes, Eugene H., author.
Hallelujah moments : tales of drug discovery / Eugene H. Cordes.
    p.   ; cm.
Includes bibliographical references.
ISBN 978-0-19-933714-9 (alk. paper)
I. Title. [DNLM: 1. Drug Discovery—history. 2. Chemistry, Pharmaceutical—methods.
3. History, 20th Century. 4. History, 21st Century.
5. Pharmaceutical Preparations—history. QV 711.1]
RM301.25
615.1'9—dc23   2013023203

9 8 7 6 5 4 3 2 1

Printed in the United States of America
on acid-free paper

Dedicated to the memory of Ralph Franz Hirschmann

Contents

Preface

TWO THINGS ARE absolutely clear. First, science and its synergistic partner technology are really important in our daily lives, now and going forward. Second, the general population, both in the United States and abroad, suffers from a high degree of scientific illiteracy. As a general rule, we are ignorant of what is really important to all of us.

It is much easier to list things for which science and technology matter critically than to list things for which they do not. Science and technology are at the heart of human and animal health, agricultural productivity, climate change, communications, national defense, economic progress, personal wealth, travel, space exploration, environmental protection, biodiversity, understanding of human history, protection from the elements, and on and on. I will leave it to you to list matters that affect human well-being that are somehow independent of science and technology.

All the matters listed in the previous paragraph are fodder for policymaking. How much of the national treasure should be invested in our science-based federal organizations, including the National Institutes of Health and the National Science Foundation? How do we handle genetically modified foods to ensure both human health and economic growth? What should we be doing to ameliorate global warning and prepare for its consequences? How should we invest to ensure that we have access to clean water and clean air? How do we best protect the abundant biodiversity that our planet offers? What investments in advanced weapons systems will provide maximal national protection at tolerable cost? Should we invest more national treasure in space exploration and, if so, how?

These and related issues are frequently areas of national contention. Given that reasonable and informed people will legitimately disagree on priorities and prospects, it is also true that much of the contention derives from ignorance. Were we, as a people, better informed about science, many of the arguments would fade, better policy decisions would be made, and we would all be better off for it. The people concerned most directly with national policymaking—our senators and representatives—are no better informed on scientific issues than the general public. And our state senators and representatives are no improvement over those at the federal level.

Scientific illiteracy leads to all sorts of problems. Almost any change will lead some people to perceive themselves as losers as a result of the change. They will fight back with whatever means they have at their disposal, fair or foul. To the extent that we are ignorant of the evidence or cannot judge the validity of the evidence relating to the change accurately, we are victims of those who attack falsely to ensure their personal or corporate benefit.

I can think of no better example than that of the tobacco industry, which effectively resisted evidence of the health threat of tobacco use for years after it was clear to all informed people the threat was real and urgent. Cigarette smoking remains the largest single threat to human health. Through misdirection, misrepresentation, bad science, and creation of a sense of uncertainty where none existed, the tobacco industry protected their profits for years at the expense of a profound negative impact on human health. Our ignorance let the tobacco industry get away with it.

Such stories continue. Specious arguments about global warming, for example, are everywhere, promoted by those global-warming-deniers who see themselves as losers should effective action be taken to combat the real threat. Whether bad policy and bad decisions derive from ignorance or venality, they affect society adversely. We need to do better.

My hope is that this little book has some modest impact on scientific literacy. It seems sensible to me to write about something that I understand. I have spent half of my professional life as a professor of chemistry and biochemistry, and the other half as a scientist doing drug discovery and development in the pharmaceutical industry, so that is what I know most about.

Getting new and useful drugs in medical practice is both important to human health and is understood poorly outside of pharmaceutical houses. It is, I believe, worth knowing about. Beyond that, drug discovery is, in a very real sense, an adventure story. Climbing Mt. Everest is an undertaking in which courage and stamina battle extreme cold, excessive winds, and high altitudes. Gripping adventure stories result. So do failures. Drug discovery is an undertaking in which wit and determination battle uncertainties in human physiology, issues of molecular design, and questions of how to interpret data in the face of regulatory uncertainties. Gripping adventure stories result. So do failures. When you start an ascent of Mt. Everest, your chances of getting to the summit are really pretty good. When you start a drug discovery project, your chances of creating a successful human health product are rather dismal. Ascending Mt. Everest gives you a story to tell. Making a drug has far-reaching consequences, including a positive effect on human health worldwide, which gives you a story to tell.

The early part of this book provides background to help you understand the tales that follow. The tales are adventure stories. I hope you understand them, learn from them, and are amused by them.

One theme that affects most of these stories is the role of basic research in providing the inspiration or knowledge base for success in drug discovery and development. For example, basic studies in human genetics and hormone physiology laid the foundation for the discovery of finasteride (chapter 6); the discovery of angiotensin-converting enzyme (ACE) inhibitors (chapter 7) relied on an understanding of blood pressure control, the chemistry of snake venom peptides, and a novel enzyme inhibitor design; discovery of the statins (chapter 8) followed an understanding of the role of cholesterol in human physiology, the biosynthetic pathway to cholesterol, and elucidation of the role of the low-density lipoprotein receptor in control of cholesterol metabolism; and the discovery of sitagliptin and its relatives (chapter 12) depended on knowledge uncovered by the Human Genome Project.

This basic scientific information was developed, in great measure, in academic laboratories and then exploited by the pharmaceutical industry to the benefit of people worldwide. Although basic research may seem arcane or even pointless, the fact is that it is the best investment in wealth creation and human welfare that can be imagined.

A good bit of the ground covered in the drug discovery and development chapters has been revealed in a different context in the book *Basic Research, Medicine, and Merck* by Roy Vagelos and Louis Galambos. You might wish to have a look at that book as well.

Acknowledgments

MANY PEOPLE HELPED me get these stories straight. Without their thoughtful assistance there would have been little here worth reading. I owe thanks to

William Campbell, who knows more about the avermectins than anyone else on the planet. His insights and patience were essential as I pulled together the avermectin story. Without his guidance, there would have been no avermectin tale in these pages.

Fred Kahan, whose help was critical to two tales in this book: Primaxin and fludalanine. He was involved intimately in both programs. His memory of events after 30 years is remarkable. Primaxin would not have been a drug without Fred's work, and fludalanine nearly made it because of it.

Art Patchett and Eve Slater, both of great help in pulling together both the ACE inhibitor and statin stories.

Nancy Thornberry, Ann Weber, and Keith Kaufmann, who revealed the Januvia/ Janumet story to me patiently and thoughtfully.

Elizabeth Morales, who expertly executed many of the drawings in this book.

Catherine Ohala, whose many edits have improved both clarity and meaning.

The following people read some or all of the manuscript, or provided key information. Their comments and criticisms helped ensure both accuracy and clarity: Georg Albers-Schönberg, Paul Anderson, Frank Ascione, Jerome Birnbaum, Burt Christensen, Shirley

Cordes, Sue Cordes, Ed Grabowski, James MacDonald, Steve Marburg, Ara Paul, and Larry Sternson.

I am grateful to (the late) Ralph F. Hirschmann and P. Roy Vagelos, from whom I learned a great deal at Merck. I would not have known enough to write this book without their teaching. Last, I am indebted to Jack W. Frazer for his wise counsel and friendship before Merck, during Merck, and after Merck.

My sincere thanks go to all the individuals mentioned here. Whatever in this book remains unclear, incomplete, or simply wrong is my sole responsibility.

HALLELUJAH MOMENTS

1 Seduced by Drug Discovery

IN AUTUMN 1978, a gentleman scientist named Ralph Hirschmann changed my life.

At the time, Ralph was a 56-year-old chemist who had just been promoted to senior vice-president for basic research in chemistry in the Merck Research Laboratories of Merck and Company,[1] a large and—in the opinion of many—the best pharmaceutical house in the world at that time. Ralph had spent his entire professional career at Merck, starting in 1950, and he had a substantial list of scientific accomplishments to his credit. One of these stood out above all others: the laboratory synthesis of a really big molecule.

The focal point of chemistry is the molecule. Linking together atoms of the elements hydrogen, oxygen, and carbon, for example, with chemical bonds (think electron glue) creates molecules. There are a lot of different molecules on planet Earth—perhaps a hundred million—some assembled by living organisms and others made in chemistry labs. Some are very small, just two or three atoms linked together. The principal components of our atmosphere—nitrogen and oxygen—are examples. Nitrogen gas consists of two nitrogen atoms (N) linked together: N_2. Likewise, oxygen gas is composed of two oxygen atoms (O) linked together: O_2. Water provides a slightly more complex example. Two hydrogen atoms (H) are linked to an oxygen atom, H-O-H, more commonly written as H_2O. Others are really big and contain thousands of atoms. This is where Ralph comes in.

The outstanding achievement for which Ralph gained fame in the arcane world of chemistry was the total laboratory synthesis of a protein—known as *ribonuclease S*—completed in 1969. A word of warning here: chemistry is full of long words such as *ribonuclease* that are difficult to spell, difficult to pronounce, and have meaning only to a chemist. There is nothing that I can do about that, so get used to it.

Proteins are big molecules—thousands of atoms. The work that Ralph did was in collaboration with another chemist at Merck—Bob Denkewalter. Sometimes two people working together can do things that neither working alone could do. At about the same time, R. Bruce Merrifield at Rockefeller University in New York City completed the total synthesis of the closely related molecule ribonuclease A, using different technology.[2]

To grasp the importance of this work, we need a word or two here about proteins (more follows in later chapters). Proteins are products of living organisms and are found in nature only in living organisms or as products of living organisms. Proteins are molecules of life. Linking together simpler molecules known as *amino acids* into a linear chain creates proteins. You do not need to worry here about just what an amino acid is. For the moment, just think of them as building blocks of proteins. Putting amino acids together to create a protein is a little like hooking Lego pieces together to make a chain. It is a bit more difficult than you may imagine; there are 20 different amino acids in proteins and the order in which they are hooked up matters. So you would need 20 different colors of Lego pieces and need to make sure that you choose the right color for each piece you add. For many years, chemists had struggled to develop the art of doing what nature does readily—hooking up amino acids into chains in an ordered fashion to create specific proteins.

This was no easy task. The chemistry involved in hooking up amino acids one after another in a specific order demanded meticulous attention to experimental detail. The longer the chain was extended, the tougher it was to get the chemistry right. To provide perspective, Vincent du Vigneaud won the Nobel Prize in Chemistry in 1955 for the isolation, structure determination, and *synthesis* of the hormone oxytocin. This molecule of life affects certain activities of the human body. It is sometimes known as the *love hormone* for its role in social recognition, pair bonding, maternal behavior, and orgasm. Oxytocin contains a (cyclic) chain of nine amino acids. Ribonuclease S has a chain of 104 amino acids! So the synthesis of ribonuclease S was orders of magnitude more challenging than that of oxytocin. This is not to argue that du Vigneaud did not merit his Nobel Prize; he surely did. The point is that the synthesis of ribonuclease S (or ribonuclease A) was a terrific accomplishment—the chemistry analog of putting a man on the moon. Science moves forward; the technology for laboratory protein synthesis has evolved and is now routine. On many fronts, what is difficult or impossible in science at one point in time becomes manageable or even routine a few years later.

Hirschmann and Denkewalter synthesized ribonuclease S using classic methods of synthesis, not very different from the methods du Vigneaud used to make oxytocin.

Merrifield accomplished the same goal using a novel technique known as *solid-state pep-tide synthesis*, which he developed. Merrifield was awarded the Nobel Prize for Chemistry in 1984 for his work, which he certainly deserved.

The laboratory synthesis of a protein was far more than a triumph of wit and will in chemistry. To provide a different perspective for this achievement, I will retreat to my time as an undergraduate chemistry major at Caltech, 1954 to 1958. As part of my education, I carried out undergraduate research in the laboratory of Richard Schweet in the department of biology, led by George Beadle, Nobel Laureate, and, later, president of the University of Chicago. At about 3:30 each afternoon, some of the scientists in the department met in a lecture room to take a cup of tea and discuss matters scientific. On occasion, the question of whether chemists would ever succeed in the laboratory synthesis of a protein would arise and engender active debate. Many thought not. They argued that doing so involved scaling a mountain higher than chemists could climb. Perhaps there were some feats of chemical synthesis that were the province of nature alone. Nature is a really terrific chemist, routinely creating molecules, including proteins, of amazing beauty and complexity. But not everyone agreed.

The most vocal advocate for the eventual triumph of chemistry for protein synthesis was a brash graduate student named Howard Temin who asserted that the probability of success was unity—that is to say, a certainty. Beyond that, he thought that success would have profound effects on organized religion. What would be the impact of humans creating in the laboratory those molecules that lie at the center of living organisms?

Hirschmann, Denkewalter, and Merrifield proved Temin half right. Laboratory synthesis of a protein was a done deal; organized religion took no notice. By the way, Temin had a splendid academic career and shared the Nobel Prize in Physiology or Medicine in 1975. Ralph Hirschmann won many honors for his work, including the National Medal of Science in 2000.[3] Bob Denkewalter failed in an attempt to organize a leadership coup within Merck Research, suffered the fate of coup leaders who fail, and took himself off to life in Alaska. He was, in any case, a superb chemist and a highly able leader, part of Merck lore.

Incidentally, after 1969, the argument about the ability of chemists to synthesize proteins has evolved into one about the ability of chemists to synthesize living organisms in the laboratory. The synthesis of proteins, DNA, and RNA—molecules that occur in all forms of life—is now routine. A complete genome, the DNA molecule that encodes all the genetic information of an organism, of a novel organism has been synthesized and inserted into the shell of a bacterium depleted of its genome. This action creates a living organism that never existed before. This accomplishment falls short of the complete laboratory creation of a novel living organism, but lends credence to our eventual capability to do so. Whether this is a good idea is an open question. I am going to don the mantle of Howard Temin here and opine that chemists will get the job done. It is up to society as a whole to ensure that the ability to create novel forms of life is put to appropriate use. And then there is the Temin question: what will be the impact of total laboratory synthesis of novel living organisms on religion? In my view, not much will happen this time, either. However, people everywhere are going to be concerned, as they should be, about the potential for good and evil in this technology. There is basically nothing that one person creates for the purpose of doing good that another person will not use for purposes of evil. Now let's get back to my story.

Merck Beckoned and I Accepted

In autumn 1978, Ralph Hirschmann showed up at the Department of Chemistry at Indiana University in Bloomington to give an afternoon seminar. This meeting was a regular affair in the department: science at 4 p.m. each Friday, usually presented by an invited visitor whose work was of interest to the department. A cocktail party often followed at one of the faculty homes—all quite civilized. At the time, I was a professor of chemistry at Indiana and had just completed 6 years as chairman of the department. It was my task to introduce Ralph and his seminar topic. He proceeded to give a meticulously prepared seminar on some aspects of his research, surely one of the best talks of the year. Ralph prepared his talks with the same level of attention to detail that he devoted to his chemistry. I later learned that everything that Ralph did was done painstakingly. Ralph's early schooling took place in Germany and, as one of my Merck colleagues pointed out, "In the Gymnasium [secondary school], they taught Ralph to pay attention to detail." Ralph had taken the lesson to heart.

Ralph had two reasons for visiting Indiana University: the obvious one was to give a seminar and enhance the reputation of Merck Research in the academic community, which actually required little enhancement. The unobvious reason was to sound me out as a possible head of a new biochemistry department at Merck. He did this in some detail as I drove him to Indianapolis after his talk, where a chemistry professor from Purdue University met and carried him on to West Lafayette for a visit there.

En route, Ralph proposed that I visit the Rahway, New Jersey, site of Merck Research for an interview. Rahway is a blue-collar town in the urban sprawl of New Jersey across the way from the bigger urban sprawl of New York City, and *nothing* like the academic/rustic charm of Bloomington, Indiana. Anyway, I agreed to do so sometime during the next few weeks, thinking that it would be interesting to see what Merck scientists were up to and to enjoy the hospitality of a highly profitable pharmaceutical enterprise.

There was good reason to be interested in Merck Research in 1978 (and there still is). A few years earlier, Roy Vagelos, one of the most prominent of U.S. academic biochemists, had left his professorial position at Washington University in Saint Louis, Missouri, to join Merck as president of Merck Research. At the time, movement of academic scientists with secure positions to industry was quite rare. Success in the academic world comes with a grant of tenure. Unless you behave very badly, you have a job for life. Industry offers no such guarantee, as many ex-employees will testify. Roy's move attracted considerable attention, mine included. The great strength of Merck Research had always been chemistry; Merck was very, very good at it (and still is). The biological sciences, although competent, were not up to the same standard at Merck. Roy Vagelos, who was a medical doctor by training and a biochemist in practice, joined Merck with the intent of getting the sciences in balance there by strengthening the biology side while maintaining the traditional strength in chemistry. So there was clearly opportunity at Merck for scientists like myself who were practicing biochemists, straddling chemistry and biology. Anyway, I had an open mind about Merck Research before my visit.

The movement of academic scientists to industry has had mixed results. Some make the transition easily and are successful in their new positions; others find it difficult to adjust to a new and quite different environment. In research-intensive universities, professors basically do not have a boss. There are department heads and deans, of course, but they matter less than you might imagine for the day-to-day activities of research-oriented professors. For the most part, the profs set their own goals, raise the resources required to do their science (largely through government grants), and depend on the university to provide an environment in which their work can succeed. Department heads and deans have resources and allocate them, and that is important, but they do not tell researchers what to do. The products of professorial research are publications in scientific journals. Their motivation is largely to satisfy innate curiosity about how the world works— gaining applause from their academic colleagues helps, as does a living wage and job security. The industrial scientific community follows academic science closely because basic research opens avenues for commercial exploitation.

In contrast, management sets sweeping goals in industrial science, leaving scientists to navigate the obstacle course set by nature, separating them from their objectives. And then, of course, in industry, one has a boss. In industry, the boss matters a good bit more than a department head or a dean does in the life of a research-oriented professor. There are adjustments to be made in moving from academe to industry. Roy Vagelos did a terrific job in effecting this change. His tenure as president of Merck Research Laboratories was a highly successful period during which good science was translated into life-saving pharmaceuticals.[4] Telling some of the stories of creating those pharmaceuticals is the central focus of this book. Merck was named The Most Admired Company in the United States for 7 consecutive years during his tenure. Roy eventually became the chief executive officer (CEO) and president of Merck and Company, and an industry icon.

Back to my story. At dinner, Ralph and I talked about science at Merck for a couple of hours. His enthusiasm for science in general and Merck research in particular was evident, and it was a joy to discuss it with him.

Life was genuinely interesting the next day. My interview trip to Merck more than lived up to my expectations. I found the science associated with drug discovery exciting, and the scientists associated with it capable and committed. Beyond that, Roy Vagelos had committed to a reorganized and expanded biochemistry department. A new laboratory building was near completion and the second floor was available to house the expansion in biochemistry. And I was a candidate to head the department—quite a head-turning proposition.

Beyond that, there was an issue associated with seeing the fruits of one's research labors realized in one's lifetime. Basic research in science provides an enormous return on investment—the best return that one ever hopes to realize. For example, the value of the annual hybrid corn crop in the United States is greater than the sum of all expenditures ever made to support research in genetics. However, there is a long lag time between the investment and the return, and it is, in general, difficult to trace back the key research-derived insights that generated the return. In industry, in contrast, one can hope to play a useful and visible role in creating something of benefit to society during the course of a career. In the pharmaceutical industry (sometimes referred to as *pharma industry*), that usually takes the form of a new drug to meet a medical need—say, a drug to treat some form of cancer; or to prevent heart attacks or strokes; or to cure infections, relieve pain, brighten mood; and on and on. That opportunity appealed to me. Beyond that, science in industry does involve an element of basic research. The total laboratory synthesis of ribonuclease S was not intended to treat or prevent a disease; it was intended to create in-house scientific expertise useful in drug discovery, and it met that goal. Merck developed a world-class peptide (think very small protein) synthesis unit at its West Point, Pennsylvania, site.

After a second interview at Merck, I was offered a position as executive director of biochemistry in Merck Research, and I accepted. It worked out well. I was happy at Merck; Merck was happy with me. Eventually, I was promoted to vice-president for biochemistry and molecular biology. I had major responsibilities in drug metabolism and drug delivery, and had a group of 400 scientists in my care. I had been seduced away from the academic world and into the pharmaceutical industry by the scientific excitement associated with drug discovery.

I want to make one thing clear. Ralph Hirschmann was not only a splendid chemist but also a great man to work for. He had vast experience in the pharmaceutical industry and willingly shared his knowledge and insights. He was a good coach and a patient man. Personally, Ralph looked and acted a lot like an academic type, except that he dressed better and would enjoy a larger retirement package. I learned a great deal from him and will always be grateful to him.

This Book Tells Tales of Drug Discovery

The objective of this book is to tell you something about how drugs for human health are discovered, developed, and brought into clinical use to diagnose, prevent, and cure human diseases. I am going to do this by telling stories of drug discoveries at Merck, most of them from when I worked there. There are a lot of these stories in the pharma industry. I have chosen a set of those at Merck for several reasons. First, these stories reflect the thinking and the work involved in getting an idea from the laboratory into clinical practice. Although technologies useful in doing this have surely changed, the fundamentals have not. Second, I was there at the time and either participated in these discovery efforts or witnessed them up close and personal (with the exception of the sitagliptin story in chapter 12). Third, the experts in each drug discovery effort are still around and available to me. Talking to them helps to ensure that I get both the details and the sense of each project right. Last, each of these stories is exciting in its own way. I hope to transmit some of the excitement to you.

I wrote this introductory blurb about my migration to Merck Research because it opened my eyes to a different and exciting world of drug discovery. My job—through the stories I relate in this book—is to do the same for you. There are seven tales, six of which led to new drugs in clinical practice that meet a significant medical need in metabolic, cardiovascular, and infectious diseases. The seventh story tells the tale of a failure, but a most interesting one. I believe that it is as close as one could ever come to an amazing success without quite getting there. I hope that you enjoy these stories and come to share the sense of adventure in them.

Drug discovery and development is a little like trying to ride a bicycle from Maine to Death Valley knowing that many of the roads along the way are closed and many of the bridges are out, and you don't know which ones. The trip is long, exhausting, filled with dead ends, and you may or may not get there at all. Two big differences are that drug discovery costs a lot more and takes a lot more time than biking across the country. Beyond that, drug discovery and development takes 100 to 200 people working together to arrive at the end point, whereas biking to Death Valley can be done alone.

People are too optimistic. No one gets married thinking it will end in divorce, but about half of all marriages in the United States end that way. No one starts a business believing it will fail, but most new businesses do fail. I think that the extreme example of overoptimism is that of discovery scientists in the pharma industry. Almost everything—say 99 percent—that pharmaceutical scientists start fails at some point. So you might well imagine a pervasive sense of gloom in anticipation of almost certain failure at the outset of a new drug discovery effort. Never happens. Excitement reigns. People are too optimistic. They persevere. And Nancy Thornberry, one of the heroes of the last story in this book, explained why: "Discovering an important new medicine is the goal of every person who works in pharmaceutical research. Until it actually happens, though, there is no way to know how absolutely thrilling it is, and how incredibly and deeply satisfying it feels."[5] Our too-optimistic people are seeking out the thrill and satisfaction.

Nancy's words also explain why I chose *Hallelujah Moments* as the title of this book. After so much failure, a success demands a celebration. Hallelujah! So there are six hallelujah moments in this book and one close-but-no-hallelujah. Everyone should shout, "Hallelujah!" when an important new drug enters clinical practice, because the world is better off for it.

This Book Is Organized: Here Is How

Having gotten through this introductory chapter, the remainder of this book falls into several parts. Chapter 2 provides a minimal chemistry background for those of you who never knew about chemistry or who have forgotten it if you did. There should be enough chemistry insights here to enable you to understand what follows. If you have a chemistry background, there is little reason to spend time on chapter 2. Chapters 3 and 4 are devoted to the chemistry of proteins. Proteins are the molecular targets of each tale of drug discovery that follows, so the essentials of protein chemistry are key to understanding the tales. If you have a good background in biochemistry, there is little reason to spend time on these chapters. Chapter 5 describes the essentials of drug discovery and development. Getting your mind around the process and the vocabulary is important to understand the drug discovery stories. Chapters 6 through 12 tell the tales: one tale to each chapter.

Here is a word of warning. The tales of drug discovery contain a few chemical structures. There are not very many, but there are some. Many of you may have experience in organic chemistry, and these structures will make sense to you. Beyond that, the structures will enrich the tales for you, so it makes sense to include them. However, the stories make sense even if these structures have no meaning for you at all. Just ignore them, understand the storyline, and move on—little lost and never mind. If you are on your way to a good restaurant and your car hits a pothole, you do not stop and go home. Think of these structures as potholes.

So get on your bike and start peddling. We have a long way to travel. We are in the piney woods of Maine; Death Valley is a continent away and the way is not yet clear.

2 The Small Molecules of Life

ALL LIFE ON Earth is unified. Life may have flickered into being, only to be subsequently extinguished, many times during the early days of our planet's evolution. But on exactly one occasion, life on Earth did arise and persist. Every living organism is a descendent of that life. We are all hatched from the same primeval egg.

The universal roles of the big molecules of life—proteins and nucleic acids—reflect this unique origin. The genetic code that links the language of nucleic acids to that of proteins is universal throughout the amazing diversity of living organisms. Protein relatives serve the same or similar functions in living organisms from wheat to humans. We are going to have a closer look at the proteins as we move forward: protein structure in chapter 3, protein function in chapter 4, and proteins as targets for drug discovery in chapters 6 through 12.

This is not to argue that there are no differences among the molecules of life. Clearly, there are. For example, bacteria are isolated from their environment by a surrounding cell wall. There is no related structure in mammalian cells. We take advantage of these differences to sustain and restore human health. For example, many antibiotics act by preventing construction of bacterial cell walls. We will see two examples in what follows: Primaxin and fludalanine.[1]

The unity of life extends to the small molecules of life as well. There is compelling similarity among the small molecules that carry out critical functions of life. Adenosine triphosphate (ATP) is the universal energy currency of life; molecules that transmit messages from one nerve cell to another are shared between sea snails and humans.[2] Molecules on the routes of metabolic pathways are much the same in fruit flies and flying bats, and on and on.

Small molecules are the topic for this chapter. We need to gain a feel for these constructs of modest size, beginning with the smallest. Later, we explore the nature of the interactions of small molecules with large ones—the basis of many phenomena of life, including adaptation to changing environments, taste, smell, and the action of most drugs used in clinical medicine, with examples to follow. Vitamins are small molecules. Small molecules carry the fragrance of flowers and the aromas of fine wine and baking bread. Small molecules provide the colors of spring blossoms and fall foliage.

To get started on this, I am going to take a short timeout here to say a few words about the nature of molecules.

Linking Atoms Using Electrons Makes Molecules

Basically, an atom is a pretty simple thing. Each atom has a really tiny nucleus, about 0.0000000001 m (or 1×10^{-10} m) across, in which a variable number of protons and neutrons are assembled. Surrounding the nucleus, there is a cloud of electrons (Figure 2.1). The number of electrons in the cloud is equal to the number of protons in the nucleus. The electron cloud is about 1,000 times larger in diameter than the atomic nucleus. So, an atom is a structure about 0.0000001 m (or 1×10^{-7} m) across—very

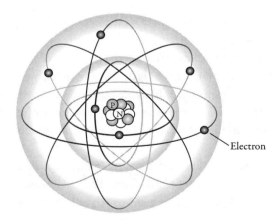

FIGURE 2.1 The Carbon Atom. Here is a highly stylized image of an atom—a carbon atom, in this case—with six protons (P) in the nucleus and six electrons in the cloud. The dark circle in the center of the image denotes the atomic nucleus, the home of the protons and neutrons (N). The ellipses surrounding the nucleus symbolize the electron cloud that surrounds the nucleus. The number of electrons in the cloud is equal to the number of protons in the nucleus. In this image, the size of the nucleus in relation to the size of the electron cloud is greatly exaggerated. In reality, the diameter of the atom is about 1,000 times greater than the diameter of the atomic nucleus.

small, as I illustrate later. Protons carry a very small positive charge. Electrons carry a negative charge equal in size to the charge of the proton. Because the number of protons and electrons in an atom is the same, the atom is electrically neutral. Neutrons are uncharged.

An element is a substance composed of a single type of atom. Elements cannot be broken down into simpler substances. Containing just one type of atom, they are as simple as it gets. Familiar examples include the metals iron, nickel, tin, lead, and mercury. An iron rod, for example, is composed entirely of iron atoms, each with 26 protons in the nucleus.

There are 92 elements that occur in nature and a bunch more that physicists have made in large and powerful machines. The nature of the element depends on the number of protons in the atomic nucleus. If there is just one, we have hydrogen; if two, we have helium; if six, we have carbon; if eight, we have oxygen; if 26, we have iron; if 92, we have uranium, and so on. Scientists find it necessary and useful to give each element a one- or two-letter designation: H stands for hydrogen, He for helium, C for carbon, O for oxygen, Fe for iron, U for uranium, and so on. That is all we need to know about atoms going forward.

The next point is that molecules are simply collections of atoms held together by chemical bonds. The chemical bonds are generally composed of pairs of electrons. A straight line, —, designates a chemical bond. The line stands for the electronic glue that holds atoms together in molecules. Chemical bonds are usually strong. It requires energy to break them. Thus, most molecules are stable.

Let's think about a couple of examples. Suppose that we take two hydrogen atoms, H and H, and one oxygen atom, O. Now let's make a chemical bond between each H and the O. We get H-O-H, the water molecule that is more conventionally written as H_2O (Figure 2.2).

We have taken three atoms, linked them together with two chemical bonds, and created a molecule. You cannot just do this willy-nilly; there are rules. For example, an alternative way of linking our three atoms together can be written as H-H-O, which turns out to violate the rules. For example, one rule is that oxygen usually makes two chemical bonds. That rule fits H-O-H, but not H-H-O, in which the oxygen atom makes only one chemical bond. Another rule is that hydrogen makes only one chemical bond. That rule also fits H-O-H, but not H-H-O, in which the central hydrogen atom makes two chemical bonds. If you want to understand all the rules, read a chemistry textbook. For our purposes, they do not much matter. What matters is that linking atoms together with chemical bonds makes molecules.

$$H + H + O \rightarrow H\text{---}O\text{---}H$$

FIGURE 2.2 Linking Hydrogen and Oxygen Atoms Creates a Water Molecule. Two hydrogen atoms combine with one oxygen atom to form water, a molecule with a central oxygen atom linked to both hydrogen atoms. The water molecule is more commonly written as H_2O.

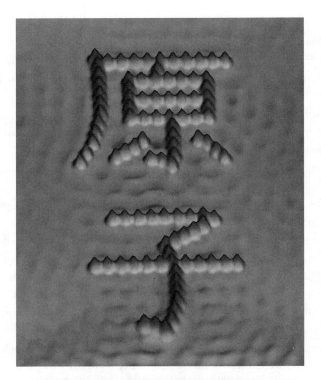

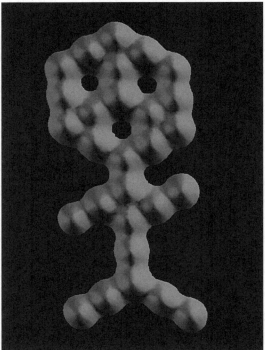

FIGURE 2.3 Individual Atoms and Molecules Can Be Manipulated. (A) Assembling atoms on a metal surface creates the Japanese characters for the word *atom*. (B) Molecules of carbon monoxide, CO, create a cartoon of a man. Both images were generated by the technique of scanning tunneling microscopy and are provided courtesy of IBM.

Molecules are really small. To get the idea, let's ask the question: how many water molecules are there in a bit more than half an ounce (actually, 18 g) of water? The answer is 60,220,000,000,000,000,000,000,000! It is easier to write this number in scientific notation[3]: 6.022×10^{23}. This is a really big number. Here are two ways to understand it.

The small chocolate candies known as M&M's provide one way to think about the size of the number 6×10^{23}. That number of M&M's would cover the lower 48 states of the United States at a depth of roughly 50 miles!

Time provides another dimension. Think about 6×10^{23} seconds. This turns out to be about 2×10^{16} years (or 20,000,000,000,000,000 years or 20 million billion years). It has been about 2,000 years since the birth of Christ. So 6×10^{23} seconds is 10,000 billion times (10^{13}) as long as the time from the birth of Christ until the present. The Earth is about 4.6 billion (4.6×10^9) years old. So 6×10^{23} seconds is five million times longer than the age of the Earth (and about 2 million times longer than the age of the universe). There are a lot of water molecules in half an ounce of water! The number of water molecules in a drop of water is approximately equal to the number of grains of sand on all the beaches on Earth. A molecule of water is really, really small.

As small as atoms are, they can be visualized individually and manipulated by a technique known as *scanning tunneling microscopy*. Have a look at Figure 2.3. In view A, the Japanese characters for the word *atom* are spelled out by atoms. In view B, arranging molecules of carbon monoxide, CO, creates a cartoon of a man.

Water Is a Truly Amazing Substance

Astrobiologists search for evidence of life beyond the borders of Earth. They face the issue of defining the absolute essentials that must be present for life to exist. There are at least three: a source of energy, a source of carbon, and water. Given all that we know about life on Earth, it is difficult to imagine that life can exist elsewhere without these three essentials.

Water lies at the very heart of life. Oceans, seas, rivers, lakes, ponds, streams, intertidal pools, and puddles provide the environment for an abundance of living organisms. Those of us who reside on land are composed largely of water and are utterly dependent on water for life. Some species can do without water longer than others, but—sooner or later—we all need it. Water has an amazing collection of unexpected properties that prove to be ideally suited for the support of life. Let's consider a few of them.

The first surprise is that water is a liquid at room temperature. Water is a very small molecule, with just three atoms: H_2O. Such small molecules are usually volatile gases at room temperature. Consider the related molecules methane (CH_4), ammonia (NH_3), and hydrogen sulfide (H_2S). All these substances are gases at room temperature: methane boils at −161°C (−258°F), ammonia at −33°C (−27°F), and hydrogen sulfide at −100°C (−148°F). Yet, water boils at 100°C (212° F). Below 0°C (32°F), water is solid, but its relatives are all gases. Two things are clear: this property of water is (1) unexpected and (2) absolutely essential for life.

Figure 2.4 provides various images of methane, ammonia, and hydrogen sulfide. This figure should help solidify the sense of the nature of small molecules.

A second unexpected property of water is that it expands when it freezes. Water has its maximal density, mass per unit volume, at 4°C. As it is cooled further, it begins to expand. Ice at 0°C (32°F) occupies about 11 percent more volume than liquid water at the same temperature. In this respect, water is nearly unique. Almost all other liquids contract when they freeze, as we would expect, because solids are generally more compact and more ordered than the corresponding liquids and, hence, are denser. This behavior is not just a laboratory curiosity. The fact is, our life on this planet is dependent on this remarkable property. This point was stated elegantly by Lawrence J. Henderson, a leading biochemist during the early 20th century, in his thoughtful book *The Fitness of the Environment*, which he wrote in 1913[4]. Here are his words:

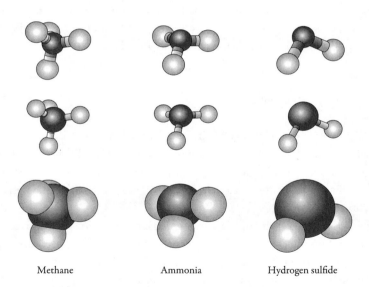

Methane   Ammonia   Hydrogen sulfide

FIGURE 2.4 Models of Simple Molecules. In the top row are ball-and-stick models of methane (CH_4), ammonia (NH_3), and hydrogen sulfide (H_2S). In all the structures, hydrogen atoms are shown in white. The central atoms—carbon, nitrogen, and sulfur, to which the hydrogen atoms are attached by chemical bonds—are shown in gray. All atoms are shown in the same size. This is not realistic, but it is perhaps the easiest way to understand the chemical bonding. In the second row are ball-and-stick models of methane, ammonia, and hydrogen sulfide in which the relative sizes of atoms are depicted. The atoms of carbon, nitrogen, and sulfur are significantly larger than those of hydrogen. In the bottom row are space-filling models of methane, ammonia, and hydrogen sulfide. The chemical bonds linking the atoms are no longer visible. The relative sizes of the various atoms are preserved. These models are the best approximations to what the molecules actually look like.

And so it would be with lakes, streams, and oceans were it not for the anomaly and the bouyance of ice. The coldest water would continually sink to the bottom and freeze there. The ice, once formed, could not be melted, because the warmer water would stay at the surface. Year after year the ice would increase in winter and persist through the summer, until eventually all or much of the body of water, according to the locality, would be turned to ice. As it is, the temperature of the bottom of a body of fresh water cannot be below the point of maximum density; on cooling further, the water rises; and ice forms only on the surface. In this way the liquid water below is protected from further cooling, and the body of water persists. In the Spring, the first warm weather melts the ice, and at the earliest possible moment all ice vanishes. So it is critical that ice floats on water, a consequence of its expansion below 4°C and, especially, when it freezes.

The facts that (1) water is a liquid at room temperature and that (2) it expands on freezing are essential to life on this planet at least. Let's turn to carbon, the element on which most of the molecules of life are based.

Elemental Carbon Occurs in Multiple Forms

Carbon is the most remarkable of elements. It combines with itself to form the framework for chains, rings, chains of rings, chains hooked onto rings, rings fused with rings, rings that penetrate other rings, various complex geometric solids, and on and on. The bonding of carbon atoms to those of hydrogen, oxygen, nitrogen, and sulfur forms most of the molecules of life. Let's get started with carbon as it occurs as the element.

Diamond is a common, if expensive, form of elemental carbon. In diamond, each carbon atom is bonded to and surrounded by four other carbon atoms located at the corners of a regular tetrahedron as shown in Figure 2.5A. Carbon atoms are highly comfortable in this particular geometry, underlying the fact that diamond is the hardest substance known that occurs in nature, a 10 on the Mohs scale.

Contrast the structure of diamond with that of graphite, another form of elemental carbon (Figure 2.5B). In graphite, the carbon atoms are organized into interconnected hexagonal rings that form extended sheets. In the sheet, each carbon atom is bonded to three others. These sheets are bound loosely to each other. The softness of graphite, a 1.5 on the Mohs scale, reflects the ability of these sheets to slide with respect to each other, and for its utility as pencil lead. However, the individual sheets of carbon atoms are very strong.

There is a central lesson here: diamond and graphite have very different properties, yet both are composed solely of carbon atoms. It follows that how the atoms are arranged in space is critical to the properties of the materials that result. This lesson is true for molecules constructed from multiple elements as well.

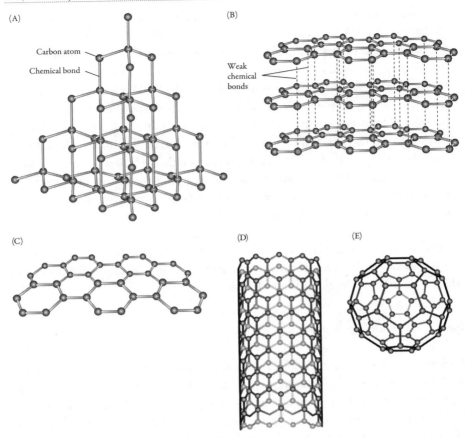

FIGURE 2.5 Forms of Elemental Carbon. (A) The basic structure of diamond. Each sphere is an atom of carbon and the straight lines connecting them are chemical bonds holding them together. This is a very small fragment of a diamond because only about 60 atoms of carbon are shown. Even the smallest diamond would contain trillions of them. (B) The structure of graphite. Here, too, the carbon atoms are shown as spheres and the solid lines are chemical bonds. The vertical dotted lines reflect weak bonds between the various layers of carbon atoms. These bonds are easily broken, which is what happens as you move the pencil tip smoothly on a piece of paper. (C) Graphene is a single layer of carbon atoms arranged in hexagons, as they are in graphite. These sheets are very tough. (D) This illustration is a nanotube. A nanotube is just a layer of graphene rolled up into a tube. (E) This is a picture of a buckyball, an enclosed figure of carbon atoms. This buckyball has 60 carbon atoms and is known as C_{60}. There are several related structures that are known that contain various numbers of carbon atoms.

A single sheet of graphite can be formed and is known as *graphene* (Figure 2.5C). Graphene is two-dimensional carbon, a sheet just one atom thick. Graphene is an exciting material: electrons move through graphene at enormous velocity. Exceedingly small transistors, containing as few as 100 carbon atoms, have been made from graphene. Efforts to pull multiple graphene transistors into a structure that would function as a computer chip are under way. Perhaps a future computer will have graphene chips rather than silicon chips. Graphene may revolutionize electronics.

A sheet of graphene can be rolled up to form a nanotube (Figure 2.5D). There are a variety of nanotube structures that differ in the subtleties of carbon atom arrangements. Nanotubes have the highest strength-to-weight ratio of any known material. They can be combined with other materials to create lightweight composites of great strength. Such composites may find use in spacecraft, for example. Nanotubes may also find use in medicine. Needlelike, they have the capacity to penetrate cell membranes into the cell interior. Chemists have attached molecules that bind preferentially to cancer cells, and anticancer drugs to nanotubes in an effort to target these drugs to cancer cells. These are still early days in searching out uses for nanotubes; their range of utilities remains to be fully explored.

Last, elemental carbon forms a variety of closed, geometric structures known generically as *fullerenes* or *buckyballs*, named after Buckminster Fuller, the American architect and designer known for his quasispherical structures. A typical example is the icosahedral (think soccer ball) structure C_{60} (Figure 2.5E). There are many sources of C_{60} in nature, including soot and stellar dust. Fullerenes, too, appear to have a number of potential applications, from storage of hydrogen to modulation of inflammation that accompanies allergic reactions.

No other element occurs in as many elemental forms as carbon. In large measure, this is true because carbon is unique in its ability to bond to itself endlessly. In all the forms of carbon illustrated in Figure 2.5, carbon bonds to itself in a variety of ways, but in highly extended structures in each case. No other element has this capacity.

Many Elements Are Found in Living Organisms

Of the 92 natural elements, quite a few are required by living organisms, including human beings. To supplement the nutrient and caloric content of my diet, I take a multiple vitamin and mineral pill each morning. In addition to the vitamins, these pills contain the following elements in varying amounts: calcium, phosphorus, iodine, magnesium, zinc, selenium, copper, manganese, chromium, molybdenum, potassium, boron, nickel, silicon, and vanadium. Some of these elements—calcium, potassium, magnesium, and phosphorus—play multiple critical roles in living systems. Others, including iodine, zinc, and manganese, have clear, well defined, if more limited roles. Some of the rest have more exotic, but essential, roles to play. Last, a few of these elements may have no essential role in human nutrition at all based on our current understanding. There they are in my vitamin and mineral pills anyway. Nature takes advantage of the properties of molecules formed from quite a number of elements in fulfilling the roles necessary to sustain life. Now let's get back to the entities that we can construct from the elements—molecules.

Here Is a Small Family of Molecules Containing Just Two Atoms

There are a substantial number of molecules formed from just two atoms, linked together by chemical bonds. Here are short portraits of four of them.

HYDROGEN, H_2, IS THE SIMPLEST POSSIBLE MOLECULE

Hydrogen is the lightest of the elements. At ordinary temperatures, hydrogen exists as a colorless diatomic gas, H_2 (sometimes called *dihydrogen* to distinguish the elemental form of hydrogen from the element itself, H). A molecule of elemental hydrogen contains two hydrogen atoms, linked together by a chemical bond: H-H.

There is very little hydrogen as H_2 in the atmosphere of the Earth, but this was not always so. There is convincing evidence that the atmosphere of the very early Earth contained much hydrogen. However, the gravitational pull of the Earth for this very light gas is weak, and atmospheric hydrogen has long since escaped into outer space.

Although there is not much hydrogen as H_2 in our atmosphere, it is the most abundant element in the universe. One simple model is that the universe is a sea of hydrogen (about 75 percent) and helium (about 25 percent) in which a few specks of other materials bob around. Closer to home, our sun is composed largely of hydrogen.

Hydrogen combines with many other elements, including carbon. Consequently, there is quite a bit of hydrogen on Earth, almost all in combination with other elements, and notably in combination with oxygen as water.

ELEMENTAL OXYGEN IS VITAL FOR MANY FORMS OF LIFE

The usual form of elemental oxygen is a diatomic gas, O_2, indispensable to life for many organisms, including human beings. Elemental oxygen as O_2 forms about 21 percent of our atmosphere, but this was not always so. In the early Earth, there was little or no oxygen in the atmosphere. About 2.7 billion years ago, the development of photosynthetic organisms, similar to the green algae that are with us today, changed that. It took another 400 million years to generate substantial atmospheric oxygen. Photosynthetic organisms take carbon dioxide and water from their surroundings and create complex molecules from these very simple ones, releasing molecular oxygen in the process. It is this oxygen that created our current, oxygen-rich atmosphere and permitted oxygen-dependent organisms, such as ourselves, to develop and thrive. You can read all about oxygen in Nick Lane's book[5] *Oxygen: The Molecule That Made the World*.

When we inhale, we take in mostly nitrogen and oxygen from our atmosphere. The nitrogen, we simply exhale; the oxygen, we use. In the lungs, oxygen attaches to molecules of hemoglobin in the red cells of the blood and is transported in the blood to the tissues. There, some oxygen detaches from the hemoglobin and diffuses into cells

(Figure 2.6). Once in the cells, it is used as a molecule that accepts electrons from metabolites derived from our food in a process that generates life-giving energy.[6]

There is a second, rarer, form of elemental oxygen: ozone, O_3, consisting of three linked oxygen atoms. Ozone is formed in the atmosphere in electrical discharges, such as lightning, and accounts for the smell of the air after a thunderstorm. Although both forms of oxygen contain only atoms of oxygen, their chemical properties are very different (Figure 2.7).

Ozone is important to us in two general ways, one positive and one not. The ozone layer in the upper atmosphere absorbs a great deal of the ultraviolet light from the sun, protecting us from damage by this high-energy light. O_2 does not have this property. The absorption of light is a sensitive function of molecular structure. O_3 absorbs ultraviolet light whereas O_2 is transparent to this radiation.

Lower in the atmosphere, ozone is a component of photochemical smog and a cause of oxidative damage to lungs. So ozone protects us in the upper atmosphere, where we are not in direct contact with it, but threatens us at ground level, where we are. We need to keep it, but keep it where it belongs.

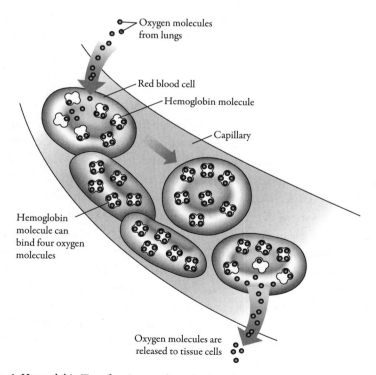

FIGURE 2.6 Hemoglobin Transfers Oxygen from the Atmosphere to Human Tissues. In the lungs, the oxygen that we inhale is, in part, attached to the protein hemoglobin in the blood. As the blood circulates through tissues, a portion of this oxygen is released to them.

O   O - O   O-O-O

Atomic oxygen   Elemental oxygen   Ozone

FIGURE 2.7 Forms of Elemental Oxygen. Oxygen (O) atoms can be linked together in different ways. Connecting two oxygen atoms with chemical bonds generates the most common form of molecular oxygen, sometimes known as *dioxygen*. Linking together three oxygen atoms creates a different molecule: ozone.

NITROGEN IS A KEY ELEMENT IN THE MOLECULES OF LIFE

Elemental nitrogen is a diatomic gas, N_2. It forms about 78 percent of our atmosphere; as noted earlier, most of the remainder is oxygen.

The characteristic chemical feature of N_2 is its stability. Vigorous conditions are required to convert N_2 into ammonia, NH_3, in the chemistry laboratory. This conversion is of great importance because ammonia is used in enormous quantities for human purposes, including fertilizer required to support intensive agriculture. In marked contrast to the conditions under which nitrogen is converted to ammonia in the laboratory, in nature, atmospheric N_2 is converted to ammonia under the mildest of conditions by enzymes called *nitrogenases* in a process termed *nitrogen fixation*. Enzymes are great facilitators of chemical reactions, increasing the rates of these reactions millions or billions of times. We return to them in chapter 4. Enzymes are also the targets of the drug discovery tales provided in chapters 6 through 12.

LAST, WHAT DON'T YOU UNDERSTAND ABOUT NO?

Oxygen and nitrogen are two of the most common elements on Earth and in living systems. As noted, both are colorless, nontoxic gases. Oxygen and nitrogen combine in a number of ways to make more complex molecules. The simplest of these is to link one nitrogen atom to one oxygen atom to yield a molecule with the composition NO, nitric oxide. Nitric oxide is a gas that is toxic in high concentrations, but, nevertheless, is a key molecule in living systems. NO has a lot of important and interesting chemistry.

Until the past couple of decades, the main interest in NO was in the context of air pollution. The basic energy-generating reaction of internal combustion engines is the explosive combination of oxygen with the hydrocarbon molecules of gasoline to produce, largely, carbon dioxide and water. The source of the oxygen is, of course, the air drawn into the engine. At the high temperatures inside the cylinders of an internal combustion engine, there is some combustion of the nitrogen—that is, some of the oxygen combines with the nitrogen in addition to the combination with the mixture of gasoline molecules. One product of that reaction is NO.

When expelled into the atmosphere as a component of exhaust gases, NO reacts with oxygen in the air to form nitrogen dioxide, NO_2, a toxic brown gas. Nitrogen dioxide, in turn, reacts with water in the atmosphere to form nitric acid, HNO_3, one of the constituents of acid rain. NO, NO_2, and HNO_3 all contribute to air pollution. These molecules are frequently lumped together under the designation NO_x. Measures taken in an effort to reduce atmospheric levels of NO_x have been successful in part, but more work remains to be done.

A related molecule is nitrous oxide, N_2O, commonly known as *laughing gas* because of its euphoric effects when inhaled. It finds some use as a recreational drug. More seriously, it is used in surgery and dentistry for its analgesic and anesthetic properties. Nitrous oxide is released into the atmosphere in substantial quantities from penguin droppings. There, it reacts with oxygen molecules to generate NO, which reacts with ozone in the stratosphere. It is a significant contributor to air pollution and is an important greenhouse gas.

More than a decade ago, it became clear that the human body makes NO. It is made in the brain, in the smooth muscle cells that form part of the lining of the blood vessels, by macrophages (white cells that form an important part of the immune system), by the corpus cavernosum of the penis, and perhaps elsewhere. NO plays an important role in each of these tissues. The lifetime of NO in the tissues is quite short, a few seconds, but it lasts long enough to be effective.

In the brain, NO appears to be involved in the reinforcement of learning, a key function. In the vasculature, it is clear that NO released by smooth muscle cells of the veins acts to relax these muscle cells, expanding the vasculature, facilitating circulation of the blood, and lowering blood pressure. Nitroglycerin and amyl nitrite have been used for many years to relieve the pain of angina pectoris, pain resulting from constriction of the coronary arteries that provide blood flow to the heart. These drugs work by releasing NO. Thus, they act to augment the amount of NO available to relax smooth muscle cells of the veins and arteries.

Macrophages are cells that fight invasion by foreign organisms. One way that they do this is to zap them with NO in concentrations that are toxic to the invaders. So NO is important as a component of the defense system of human beings. As it turns out, sodium nitrite, $NaNO_2$, a commonly used preservative, works by generating NO, which is responsible for its ability to preserve foods.

Last, erotic thoughts trigger the release of NO in the corpus cavernosum of the penis. This causes the corpus cavernosum to relax, permitting the inflow of blood, resulting in a penile erection, and the capacity for turning the erotic thoughts into erotic action. The widely prescribed drugs Viagra, Levitra, Cialis, and Stendra work by increasing the release of NO in the corpus cavernosum.

Here Are a Few Generalities About Chemistry

Chemistry gets a lot of bad ink. Many people have the impression that chemistry is the toxic science, just as some people regard economics as the dismal science. This impression gets of a lot of routine reinforcement from news sources. Chemistry is associated with pollution of the atmosphere by ozone and the noxious oxides of nitrogen and sulfur, oil spills, toxic pesticides and herbicides, smelly oil refineries, undesired food additives, chemical accidents, chemical weapons, and substances of abuse such as heroin, cocaine, and methamphetamine. Think about the use of chlorine gas and mustard gas by the German army in World War I or the use of far more sophisticated chemical weapons used by Saddam Hussein against the Kurds in northern Iraq and Shiite Muslims in southern Iraq at the end of the first Gulf War. But, chemistry is not all bad.

CHEMISTRY HAS A ROSY SIDE

Like most stories, that of chemistry has another side, less often noted or remarked on perhaps, but a whole lot more pleasant. The world of chemistry is the world of molecules. It is a complex, critical, and fascinating world. Molecules and their constituent atoms make up all matter. Specific molecules affect every aspect of our lives every day, frequently for better but occasionally for worse. The simple fact is that almost everything that we use in daily life has been modified chemically in some way: plastics, alloys, detergents and soaps, paper, perfumes and colognes, and our drinking water (fluorinated). It is difficult to imagine life without the products of chemistry.

Think about those molecules that are pharmaceutical products. They are used for multiple purposes. They treat your infections, relieve the pain of your sore throat, control your coughing, lower your blood pressure, raise your spirits, lower your plasma cholesterol, fight your allergies, keep you awake and alert, help you sleep, alleviate your heartburn, modulate your anxiety, help diagnose your disease, and render you mercifully unconscious during surgery. In serious cases, they battle cancer, improve the chances of surviving a heart attack or stroke, alleviate symptoms of schizophrenia or bipolar disorder, and reduce the incidence of suicide in severely depressed patients. The agents that contribute to human well-being are molecules.

Small molecules provide the aromas and tastes of apples, pears, apricots, kiwis, bananas, oranges, and other fruits. They are the fragrance of lilacs, roses, and fine wines, as well as those of the perfumes, colognes, after shave lotions, and deodorants that we lavish on ourselves. Wine lovers may describe a wine as smelling of raspberries, which many wines do because they share a common chemical—raspberry ketone—with raspberries themselves. The vegetative aroma of sauvignon blanc reflects a molecule—known technically as 2-methoxy-3-isobutylpyrazine—also found in bell peppers. Benzaldehyde is found in Pinot Noir—a wine frequently described as having nuances of cherries, the aroma of which also contains benzaldehyde. The brilliant colors that we enjoy in spring flowers and fall foliage—reds, yellows, oranges—are the colors of molecules. The complex molecular photosynthetic machinery of green plants and certain algae capture the radiant energy of the sun, which is ultimately responsible for almost all life on Earth.

Small Molecules Serve Multiple Purposes in Life

Earlier in this chapter, we saw multiple examples of the roles of small molecules in life: oxygen in our atmosphere that we breathe, water that sustains life, several roles of nitric oxide in physiology, and so on, but these examples just scratch the surface. Several additional key roles of small molecules in life are presented in the following. The list is far from comprehensive, but it does provide critical insight.

SMALL MOLECULES ARE BUILDING BLOCKS FOR BIG MOLECULES

Three classes of the big molecules of life are polymers: proteins, nucleic acids, and polysaccharides (starches and glycogen, for example). Polymers are large molecules constructed by hooking together small molecules (monomers) repeatedly. Proteins are linear chains constructed from a family of 20 different amino acids, nucleic acids are linear chains constructed from a family of four nucleotides, and polysaccharides may be linear or branched chains constructed, usually, from one or two different sugars. The amino acids, nucleotides, and sugars are all examples of small molecules. And all living organisms use the same family of amino acids, the same family of nucleotides, and the same family of sugars. The unity of life at the molecular level is compelling.

BIG MOLECULES BREAK DOWN TO SMALL MOLECULES IN LIVING ORGANISMS

Chemical reactions convert molecules into other molecules. There are a great many chemical reactions required to sustain life. Think about the digestion of food. You consume very large, complex molecules: proteins, complex carbohydrates, fats, and so on. But at the end of the day, you excrete very simple ones: water, carbon dioxide, and other small molecules as well as solid matter. So a lot of chemistry has happened along the way; it is called *metabolism*. Each metabolite is a specific molecule. The metabolism of an organism is the sum of all chemical reactions that sustain its life at some point in time. To get an idea of the complexity of metabolism, have a look at Figure 2.8. Here the individual metabolites are shown as dots, and the lines that connect them represent chemical transformations among them. There are several points to make about this illustration. First, it is impossible to show the complete metabolism of an organism in a diagram of this type. Second, metabolism differs somewhat from organism to organism, although the commonalities are far more impressive than the differences. Figure 2.8 captures metabolism that is common to most organisms. Third, the metabolism of an organism varies over time and under different conditions. For example, the metabolism of a pregnant woman is different in some ways from that of the same woman when not carrying a child. Your metabolism will not be the same in the middle of a 10K race as it is when sitting idle in front of the TV. The metabolism of a starving person is different from that of the same person when well fed. Nonetheless, even in these circumstances,

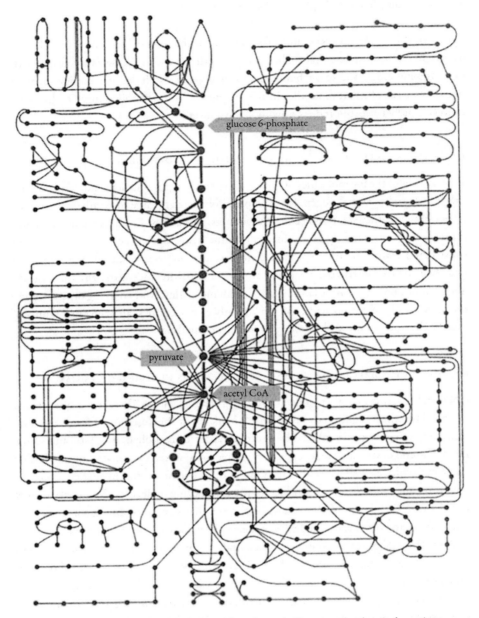

FIGURE 2.8 The Complexity of Metabolism. This schematic illustrates the chemical reactions that interconvert small molecules in cells. Each dot represents a compound; lines that connect the dots represent reactions. The heavy lines and dots down the center of the maze reflect the glycolytic pathway in which the sugar glucose is degraded to smaller molecules with production of chemical energy. The heavy circle near the bottom of the maze represents the citric acid cycle in which the small-molecule products of the glycolytic pathway are degraded to carbon dioxide and water, again with the production of chemical energy. (© 2007 From Molecular Biology of the Cell, 5th Edition, by Alberts et al. Reproduced by permission of Garland Science/Taylor and Francis, LLC.)

the similarities in metabolism far outweigh the differences. Last, metabolism may change during disease, and metabolic changes may be the basis for disease. For example, the metabolism of sugars is altered in the disease diabetes mellitus.

Each dot in Figure 2.8 represents some small molecule. The molecules may be degradation products from the digestion of large molecules (such as proteins or starches), building blocks for large molecules (perhaps proteins, nucleic acids, or glycogen), or molecules on metabolic routes that link degradative and biosynthetic pathways. Many drugs in common use to prevent or treat disease act on our metabolism in specific ways.

SMALL MOLECULES CAN CONTROL METABOLISM

As noted earlier, the metabolism of an organism changes in response to changing circumstances, which requires that there be agents that respond to changing circumstances with changes in metabolism. Many of these agents, although not all, are small molecules.

For example, many hormones are small molecules, and they are metabolic control agents. The sex hormones—androgens and estrogens—are examples, as are a number of other steroid hormones, such as corticosteroids, aldosterone, and progesterone. Derivatives of some of the vitamins, including those of vitamins A and D, are metabolic control agents. Other small molecules act as signaling molecules, intermediates that carry messages from hormones to the final site of metabolic control. Still other small molecules may act to stimulate or inhibit the activity of enzymes, the protein catalysts that function in basically all metabolic reactions.

NEUROTRANSMITTERS ARE SMALL MOLECULES

Neurotransmitters provide for chemical communication within the neural pathways in the central and peripheral nervous systems, and between the nervous system and the muscles. For example, one neuron may communicate with another across a small gap by releasing a neurotransmitter molecule that is recognized by the second one. The message is carried. These neurotransmitters are small molecules: dopamine, norepinephrine, serotonin, glycine, glutamate, and γ-aminobutyric acid. Here, too, there is commonality in life. As noted earlier, you can find some of the same neurotransmitters in sea snails and people.

VITAMINS ARE SMALL MOLECULES

Vitamins are small molecules required in the human diet. They are essential for normal physiology and cannot be synthesized by the human body (at least not in the required amounts) and so must be gained through the diet. They include riboflavin, niacin, pyridoxine, ascorbic acid (vitamin C), pantothenic acid, and vitamins A, D, E, and B_{12}. Most of these small molecules must be converted into other structures in the course of human metabolism to be active, working together with enzymes to effect necessary chemical transformations or as metabolic control agents.

MOST PHARMACEUTICALS ARE SMALL MOLECULES

Basically, all medications taken orally are small molecules. They include antibiotics; analgesics; agents to control inflammation; agents that alter mood including antidepressants, anxiolytics, and neuroleptics; molecules that lower your blood pressure or your cholesterol level, calm an upset stomach, or relieve bronchial congestion; and on and on. In addition, many pharmaceuticals given by other routes—injection, inhalation, sublingually, intranasally, by suppository—are also small molecules. Some drugs are destroyed as they travel through the gut, or they are not absorbed in the gut and so must be given by a nonoral route. In other cases, it simply makes sense to use an alternative route; drugs for asthma are frequently given by inhalation, for example. These molecules include anesthetics, antiallergens, bronchodilators, and some agents for cancer chemotherapy. Almost all medicines purchased over the counter are small molecules.

PHEROMONES ARE SMALL MOLECULES

Pheromones are molecules used to communicate between members of the same species. They are used as sex attractants, warning agents, information providers, and for many other purposes. Pheromones are used widely throughout nature, although their role in controlling human behavior remains uncertain. A number of pheromones have been identified chemically and are small molecules in various classes.

GENERATORS OF AROMAS AND TASTES ARE SMALL MOLECULES

Small molecules are responsible for the aromas of fruits, flowers, perfumes, colognes, and so on. They are also responsible for less pleasant or downright obnoxious smells such as rotten eggs, spoiled meat, or skunk spray, for example. They also account for tastes whether salty, sweet, acidic, bitter, or savory—those of fruits, vegetables, meats, candies, cakes, spices, condiments, and so on.

SMALL MOLECULES ARE DRUGS OF ABUSE

On the less happy side of the world of chemistry are drugs of abuse. Many of these small molecules have a feel-good effect coupled with addictive properties, a combination that initially encourages repeated use and later demands repeated use. Several of these drugs are opioids, molecules derived from one of the most useful drugs of all time—morphine. Morphine is an effective analgesic that is inexpensive and available worldwide. Heroin is a simple derivative of morphine and is basically a way to get a lot of morphine into the brain quickly. It is one of the primary drugs of abuse. Cocaine, methamphetamine, lysergic acid diethylamide (aka LSD), psilocybin, heroin, phencyclidine (aka PCP), and mescaline are other examples of small-molecule drugs of abuse that have a powerful influence on mental state.

Anabolic steroids fall into a distinct small-molecule class. They are not addictive and do not have potent effects on mental state. Repeated use combined with an exercise program results in improved athletic performance, and these small molecules continue to be widely abused by athletes. Professional bicycling, football, and baseball have been particularly affected by their use, although the abuse of these agents is not limited to these activities.

SMALL MOLECULES SERVE AS MARKERS FOR LIFE

As noted earlier, living organisms have been around for a long time, perhaps 3 billion years. Reconstructing the history of life on Earth is an enormously challenging task. It is amazing how much has been learned about organisms that flourished a billion or so years ago. These organisms have left chemical signatures in organic material found in rocks, deep ocean sediments, muds, and elsewhere. It has been possible to link these chemical signatures to various life forms, just as fingerprints may be linked to individuals. These chemical signatures are small molecules or families of small molecules.

SMALL MOLECULES ARE FOUND IN INTERSTELLAR SPACE

It may surprise you to know that quite a collection of small molecules have been detected and identified in interstellar space, perhaps 200 of them in total. Many of these molecules are exotic, but a fair number are familiar on Earth, including molecular hydrogen, oxygen, nitrogen, nitric acid, carbon monoxide, carbon dioxide, water, nitrous oxide, sulfur dioxide (SO_2), ammonia, methane, formic acid ($HCOOH$), formaldehyde ($HCHO$), ethylene (C_2H_4), methanol (CH_3OH), ethanol (CH_3CH_2OH), benzene (C_6H_6), and glycine (NH_2-CH_2-$COOH$), among others. Halley's Comet is known to contain the small molecules water, ammonia, hydrogen cyanide (HCN), and formaldehyde. There is more commonality between the chemistry on Earth and that in interstellar space than many of us might have imagined, which influences our chance of discovering extraterrestrial life.

I could push ahead with this list of small molecules, but my central point should be clear: small molecules are everywhere and they matter. They are the air we breathe and the water we drink, they create the aromas and taste in the food we eat, and they are the key metabolites in the chemistry of life. They are in pharmaceuticals, energy sources (natural gas, propane, butane, gasoline), colognes, aftershaves, pesticides, herbicides, and on and on.

Life depends on the interactions between small molecules and large ones. Having said something about the small ones, it is time to turn our attention to big the ones: the proteins.

3 Proteins
MOLECULAR WONDERS IN THREE DIMENSIONS

I GREW UP in a solid middle-class family, largely of German descent, in a city of modest size in central Nebraska. Like a lot of such families, our diet was based on meat and potatoes. It was an unwritten but religiously observed law in our home that two meals each day would include both meat and potatoes. The meat was turkey twice a year, ham on occasion, chicken or pork from time to time, but mostly beef. The potatoes were usually boiled or boiled potatoes subsequently sliced and fried. My brother and I also drank a lot of whole milk, at least a quart a day each and frequently more (skim milk was available, but no one gave much thought to "reduced fat" or "low fat" milk back in those days). On farms, a lot of people just drank what the cows had on tap. Between the meat, potatoes, and the whole milk, we got a lot of protein in our diet, which is good; we also got a lot of saturated fat in our diet and that is not so good.

Protein Is a Critical Nutritional Requirement

Adequate protein in our diet is essential for good health. Proteins in our diet break down to provide *essential amino acids*. Amino acids are the building blocks of proteins. The amino acids that are essential in our diet are those that our bodies cannot make or cannot make in adequate quantity for optimal health. For dietary proteins, two things matter: amount and quality. The amount of protein is a simple quantitative matter; it is measured in grams per day. The amount you need depends on several factors: your gender, age, size, level of exercise and other physical activity, and whether you are pregnant or lactating, for example. The quality of protein is not so easy to evaluate. Getting the essential amino acids in your diet is more important than getting the others. The highest quality proteins are those that contain an abundance of all the essential amino acids. Meats and dairy products are among the best sources of high-quality proteins. In many cases, these protein sources also come with rather more fat than most of us would like. Skinless chicken breast provides an attractive exception. Of course, vegetables and grains also provide a healthy source of proteins. However, it is important to eat a variety of them because individual ones may lack sufficient levels of one or more of the essential amino acids. The central lesson is clear: adequate dietary protein is important for human health.

To understand how small molecules work their biological magic as therapeutic agents that restore us to good health, for example, we need to understand some basic things about the big molecules of life, particularly the proteins. The point is that *all the wonderful things that small molecules do derives from their interactions with large molecules.* So we need to understand both the small and the large. We had a look at the small in the last chapter, so now we move on to the large.

Proteins Are Constructed from Amino Acids

Talking about big molecules does not require that we lose focus on small ones. Big molecules of life are constructed from small ones. If you link enough small molecules together you eventually get big ones. If you link enough boxcars together, you get a train—same idea. Let's see how this works for proteins.

As noted earlier, the building blocks of proteins are the amino acids. There are 20 amino acids that occur commonly in proteins. This family-of-protein building block is universal. You find the same set of 20 in life forms from vegetables to rhinos. Life at the molecular level is unified.

Proteins are critical components of all forms of life. Proteins form the coat of the simplest viruses (although whether viruses are alive is a matter of opinion). Proteins are the only detectable component of mysterious infectious particles termed *prions* (the culprits in mad cow disease). Proteins serve multiple functions in all living organisms, simple and complex. These functions include, among others, catalysis, defense, metabolic regulation, movement, and architecture (we look at catalysis in the next chapter). It is essential to know that the marvelous properties of proteins derive from their structure. So let's move on to that topic and provide the background for understanding how protein structure relates to function.

PROTEINS ARE POLYMERS OF AMINO ACIDS

Proteins are polymers. Polymers are molecules—often, very large molecules—that are made up by the repetition of one or more structural elements. Think about making long chains of Lego pieces. The simple process of hooking up unit after unit permits you to create a large structure. Other examples include making a string of beads or a paperclip chain. In each case, the underlying idea is the same: take some element or elements and hook them up in a linear fashion to create a chain.

In the case of hooking up amino acids to create a protein, this is not quite as easy as I may have implied by my analogies. There are 20 different amino acids in proteins, and the order in which they are linked matters. So we would need 20 different shapes of Lego pieces to hook together or 20 different colors of beads to string. We would need to be careful to get the order correct—blue first, then red, followed by yellow, mauve, lavender, puce, green, and so on.

Polymers occur in many of the materials you encounter in day-to-day life: polyethylene, polypropylene, polystyrene, nylon, rayon, orlon, and on and on.

The parent molecule of the amino acids is glycine (*gly-seen*): H_2N-CH_2-COOH. The NH_2 collection of atoms is known as the *amino* group. The COOH collection of atoms is known as the carboxylic *acid* group—hence we get the name amino acids. The carbon atom (C) of the CH_2 group between the amino and acid groups is called the *alpha* (α) *carbon atom*. Glycine is the simplest possible α-amino acid and is a very common constituent of proteins.

The other 19 amino acids that occur commonly in proteins can be thought of as derivatives of glycine. These are formed, with one exception, by replacing one of the hydrogen atoms (H) on the α carbon atom by some group of atoms. These groups are constructed from carbon, hydrogen, oxygen, nitrogen, and sulfur atoms.

The properties of the amino acids that result from replacing one hydrogen atom of glycine with a group of atoms depend on what that group is. There are three categories.

First, there are nine amino acids that have "greasy" replacements for the hydrogen atom. These greasy replacements consist mostly or entirely of carbon and hydrogen atoms: hydrocarbons. Basically, hydrocarbons do not like water. They are hydrophobic—meaning, water-hating. Think about fats, oils, or waxes. In living organisms, proteins exist in a watery environment. Anticipating what comes later, these greasy amino acids are going to seek some way of getting out of contact with the water in their environment.

Second, there are six amino acids with replacement groups that contain atoms of oxygen and/or nitrogen in addition to carbon and hydrogen. These amino acids are far less hydrophobic than the nine just described. In a watery environment, they are happy to be in contact with water but do not insist on it.

Last, there are five amino acids that, under physiological conditions, bear electrical charges. Three of them have positive charges and the other two have negative charges. Molecules bearing charges love water; they are strongly hydrophilic, or water-loving. In a watery environment, they insist on being in contact with water.

All of this matters for determining the three-dimensional structure of proteins, as we shall see. It also provides a way to understand the nature of several genetic (inherited) diseases such as sickle cell anemia, which we get to later.

AMINO ACIDS IN PROTEINS ARE LINKED TOGETHER

The 20 amino acids that constitute proteins are hooked together by joining the amino group of one amino acid to the carboxyl group of another with the elimination of a molecule of water (Figure 3.1):

The bond holding the amino acids together is known as a *peptide bond* (enclosed in the dashed box in Figure 3.1). Proteins are referred to frequently as *polypeptides*, a term that I use in this book from time to time.

Proteins may have a hundred, a few hundred, or a few thousand amino acids hooked together. As a starting point, we can imagine hooking two amino acids together to make something known as a *dipeptide*, three amino acids together to make a *tripeptide*, and so forth. It is easy to see where the term *polypeptide* comes from, with *poly-* meaning "many."

Here is an interesting question: how many dipeptides can we make? Well, we can have 20 different amino acids in the first position and 20 more in the second one. So for each amino acid in the first position, we can have 20 in the second. It follows that $20 \times 20 = 400$ dipeptides. How many tripeptides can we make out of our 20 amino acids? For each of the 400 dipeptides, we can have 20 different amino acids in the third position. So $400 \times 20 = 8,000$ tripeptides. There are 160,000 possible tetrapeptides (four amino acids) and 3,200,000 pentapeptides (five amino acids). The point is, that even for quite short chains of amino acids, there are a very large number of possible, different molecules, and each has its own set of properties.

If the amino acid chain is n amino acids long, the number of possible structures is 20^n. The next question is: how big is n? The fact is that n varies a lot. For insulin, quite a small protein, n is 51. More typically, n is greater than 100 and sometimes much greater. Even if we confine our attention to proteins the size of insulin, there are 20^{51} possible protein structures having 51 amino acids. This is an unimaginably large number. I could not write it down in the usual notation if I wanted to; there is not enough paper in the world on which to write this number.

Therein lies the secret of the diversity of protein functions. There are so many possible protein structures that nature, through the process of evolution, has been able to pick and choose among this cornucopia of possibilities to find the cream of the crop for each function. The number of different proteins in the human body—perhaps 100,000—is an incredibly small fraction of all the proteins one can construct using 20 natural amino acids linked in chains, say, 100 units long (one part in 10^{125}).

$$NH_2\text{-}CH_2\text{-}COOH + NH_2\text{-}CH_2\text{-}COOH \longrightarrow NH_2\text{-}CH_2\text{-}CO\text{-}NH\text{-}CH_2\text{-}COOH + H_2O$$

FIGURE 3.1 Amino Acids Are Linked by Peptide Bonds. In the example shown here, the carboxyl group of one molecule of glycine is linked to the amino group of another. The four atoms enclosed in the dashed-line box form the peptide bond that connects them.

The Sequence of Amino Acids Along the Chain Defines the Primary Structure of a Protein

So far, we know that 20 amino acids that appear in proteins are hooked together in a linear chain. We also know there are an enormous number of possible proteins.

It is clear that the most fundamental structure of a protein is defined by the sequence of amino acids along the chain. This sequence is known as the protein's *primary structure*. The primary structure demands that we specify which of the 20 amino acids comes first in the chain, which second, which third, and so on, until we reach the end of the chain.

Bovine insulin was the first protein to have its primary structure revealed; this was accomplished in 1955. This magnificent achievement was the product of the work of Frederick Sanger and his colleagues at Cambridge University in England. Sanger won the Nobel Prize in Chemistry in 1958 in recognition for this triumph of science. The primary structure of insulin is provided in Figure 3.2. Each amino acid is identified by a three-letter code. For example, glycine is denoted by Gly or gly, alanine by Ala or ala, and so forth. Insulin has two chains, held together by bridges of sulfur (S) atoms. This is unusual and reflects the way that insulin is made. Specifically, it is made in a single polypeptide chain and a segment is later cut out of the middle, leaving two chains.

The significance of Sanger's work is immense. It proved, for the first time, that the sequence of amino acids along the protein chain is unique. That is, all molecules of bovine insulin, for example, have the same sequence of amino acids along the chain. This singular finding requires that there is a genetic code. That is, information is encoded in a molecule that specifies the sequence of amino acids in the insulin molecule and, for that matter, in all protein molecules. That molecule is DNA.

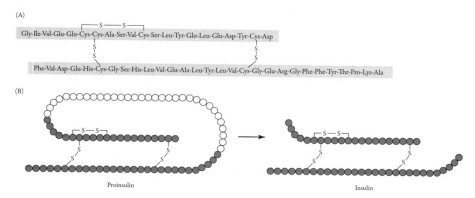

(A)

Gly-Ile-Val-Glu-Glu-Cys-Cys-Ala-Ser-Val-Cys-Ser-Leu-Tyr-Glu-Leu-Glu-Asp-Tyr-Cys-Asp

Phe-Val-Asp-Glu-His-Cys-Gly-Ser-His-Leu-Val-Glu-Ala-Leu-Tyr-Leu-Val-Cys-Gly-Glu-Arg-Gly-Phe-Phe-Tyr-Thr-Pro-Lys-Ala

(B)

Proinsulin

Insulin

FIGURE 3.2 Primary Structure of Bovine Insulin. (A) The sequence of amino acids for bovine insulin is shown, each identified by a three-letter abbreviation. This sequence is the primary structure for this hormone. Note that a pair of bridges of sulfur atoms holds the chains together. (B) This illustration shows how the two chains arise by clipping out a segment of the polypeptide synthesized initially. The initial product is known as *proinsulin*.

Ribonuclease A is a small protein that has 124 amino acids in its chain. Recall that this is the very protein first synthesized in the laboratory by Merrifield at The Rockefeller University. The sequence of amino acids for ribonuclease A—its primary structure—was determined in 1963 by Derek G. Smyth, William Stein, and Sanford Moore. It was the second protein for which the primary structure was deduced.

To provide some sense of the rate of progress in biochemistry and the rate at which information is generated, Sanger got the primary structure for insulin, a small protein, in 1955 after several years of dedicated work. We waited 8 years for the next primary structure, that for ribonuclease A, to be determined, for which a Nobel Prize in Chemistry (in part) was also awarded, in 1972, to Christian Anfinsen, Sanford Moore, and William Stein. There are now tens of thousands of such primary structures known, and the number increases substantially each and every day. The library of known primary structures for proteins is collected in major databases and is accessible to scientists or others who have need of them.

Before we take leave of primary structures for proteins, there is one last, important point to be made. Proteins serving the same function in different species may have different primary structures. For example, the primary structures of bovine, ovine, and human insulins are not quite the same. Some amino acids replace others at specific points along the chain. The resulting proteins are closely related in amino acid sequence, but they are not identical. Different species have discovered different protein solutions for the same biological problem through evolution.

One might well imagine that the primary structures for proteins serving the same function in different species would most resemble each other for species that are closely related in an evolutionary sense—such proteins are termed *homologs*—and this turns out to be the case. Cytochrome c is a protein distributed quite widely throughout nature. The primary structure for cytochrome c of humans is very similar to that of pigs, but not quite so similar to that of frogs, and is even more distantly related to that of insects and, at the extreme, that of wheat.

Extraordinarily, fragments of the protein of connective tissue known as *collagen* have been extracted from bones of the dinosaur *Tyrannosaurus rex*. The primary structure of these fragments strengthens the suggestion that birds are the closest living relatives to dinosaurs. This finding has set off a determined effort to recover additional protein material from fossils.

Proteins Have Unique Three-Dimensional Structures

When we know the primary structure—the amino acid sequence—for a protein, we have a basic aspect of the structure in hand. We know which amino acid follows which along the protein chain. Even more basically, we know which atom is bonded to which for all atoms in the protein.

We are now in a position to ask a more sophisticated question related to protein structure: what is the three-dimensional structure of the protein? This question assumes something that is by no means obvious: that a protein has a unique three-dimensional structure. Think of a protein as 100 beads on a string. You can imagine that it can fold up in a great many ways, probably approaching infinity, or not fold up at all—just like you can fold up a string of beads in a great many ways.

The following point is of great significance: *under physiological conditions, nearly all proteins have one predominant, well-defined three-dimensional structure, and the biochemical activity and biological function of that protein depends on that structure.* If the three-dimensional structure of the protein is lost, so is the biological activity. We shall see examples of this fact a bit later.

PROTEINS CONTAIN HELICES, SHEETS, AND LOOPS: SECONDARY STRUCTURE

An important element in the three-dimensional structure of a protein is known as *secondary structure*. The secondary structure results from the formation of hydrogen bonds—the sharing of a hydrogen atom between two other atoms, usually oxygen or nitrogen: -O-H. . .N-H. Hydrogen bonds are weak chemical bonds that are made and broken easily. Hydrogen bonds can be formed between different amino acids along the polypeptide chain. There are two basic ways to do this. We can form a helix or we can form a sheet. The great American chemist Linus Pauling won the Nobel Prize in Chemistry in 1954 for the elucidation of these structures.

The helical structure that occurs commonly in proteins is termed the α helix. A model of the α helix is provided in Figure 3.3 and a model of the β sheet is provided in Figure 3.4.

Helical and sheet structures in proteins may be interspersed with loops. Individual proteins vary greatly in their content of helical and sheet structures. Some are mostly helical, some mostly sheet, some contain a good deal of both, and others contain little of either. Nature provides all possibilities for our enjoyment.

THE POSITION OF EACH PROTEIN ATOM IN SPACE DEFINES ITS
THREE-DIMENSIONAL STRUCTURE: TERTIARY STRUCTURE

Consider a protein composed of a single amino acid chain. The tertiary structure defines the position of each atom in three-dimensional space. Thus, the tertiary structure includes secondary structural elements, helices and sheets, as well as the spatial distribution of the amino acid side chains of these elements and the positions of each atom in the loops between them.

Myoglobin is the first protein for which we had knowledge of protein tertiary structure. This monumental accomplishment was the product of the work of two crystallographers: John Kendrew, who actually completed the myoglobin structure in 1959, and Max Perutz, Kendrew's mentor. Perutz initiated work on the structure for hemoglobin a

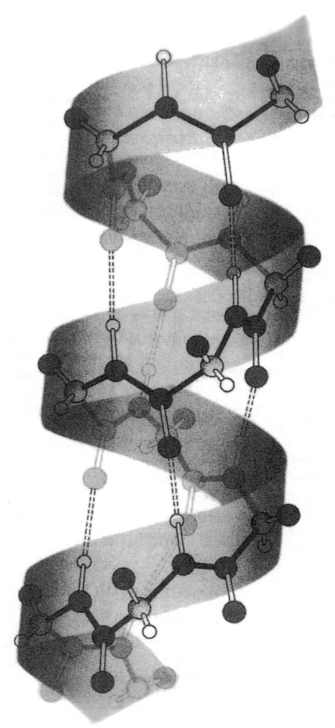

FIGURE 3.3 The Structure of the α Helix of Proteins. The main course of the helix is shown by the shaded ribbon. Hydrogen bonds between the N-H and the C=O groups along the polypeptide chain are indicated by dashed lines. There is an amino acid every 1.5 Å along the helical axis. The distance along the axis required for one turn is 5.3 Å, giving 3.6 amino acids per turn. (Illustration, Irving Geis/ Geis Archive Trust. Copyright Howard Hughes Medical Institute. Reproduced with permission.)

(A) Antiparallel

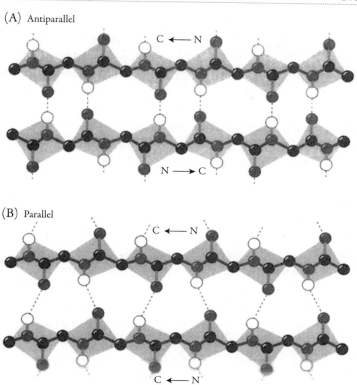

(B) Parallel

FIGURE 3.4 The β Sheet Structure of Proteins. Polypeptide chains are held together by hydrogen bonds, indicated by dashed lines, in configurations called the (A) antiparallel β pleated sheet, and (B) parallel β pleated sheet. (Illustration, Irving Geis/Geis Archive Trust. Copyright Howard Hughes Medical Institute. Reproduced with permission.)

few years earlier than Kendrew got started on myoglobin. However, myoglobin is the simpler molecule and Kendrew completed the myoglobin work a few years ahead of Perutz's work on hemoglobin. Elucidation of the structure of myoglobin and hemoglobin capped three decades of scientific effort. Kendrew and Perutz shared the Nobel Prize in Chemistry in 1962 in recognition of their accomplishments.

Myoglobin is a protein found in mammalian muscle. It is an oxygen-storing protein of modest size, containing a single chain of 153 amino acids. Myoglobin contains a single atom of iron, bound in a complex structure termed *heme* and to which a single molecule of oxygen binds. In humans, myoglobin is found in highest concentration in the heart muscle. We might have guessed this because we want to keep the heart supplied with a reserve of oxygen. The most abundant sources of myoglobin come from the muscles of diving mammals, whales and dolphins, whose myoglobin-linked stores of oxygen permit them to submerge for substantial periods of time.

We should not overlook one central finding established by the structure of myoglobin. Just as the primary sequence of each myoglobin molecule from a single species is the same, so is the three-dimensional structure. Each polypeptide chain organizes itself in space in just the same way. This is highly important because, as noted earlier, the personality of the protein is encoded in its three-dimensional structure. If the three-dimensional structure is perturbed in some way, the biological properties of the protein are likely to be altered or perhaps lost entirely. Enzymes lose catalytic activity, antibodies may no longer recognize the antigen that elicited them, receptors may fail to recognize their ligand, myoglobin may fail to bind and store oxygen, and so forth.

Here, too, we have an example of the rate at which scientific information is created currently. The earliest three-dimensional protein structures were products of decades of work. During the succeeding decades, we have defined three-dimensional structures for tens of thousands of proteins and complexes of proteins with small molecules. They occupy huge publicly available databases. These databases grow daily. This information is of great importance for several purposes. Among them is the design of novel molecules (drugs) useful in human clinical medicine. We will get to some of the case histories later in this book.

Chemists have adopted several models for depicting protein structure, including depiction of all the heavy atoms of the protein (the hydrogen atoms are deleted to keep the model simple enough to be understandable), emphasis on the backbone formed by the α carbon atoms, and ribbon diagrams that illustrate secondary structural elements. Let's make these comments concrete by looking at an example. The enzyme human renin is a player, although not the star, in one of the tales that follow—the discovery of angiotensin-converting enzyme (ACE) inhibitors. Renin is an enzyme (much more about enzymes follows in the next chapter) secreted by the kidneys and is involved critically in the control of blood pressure. It is a protein created by linking 340 amino acids together in a chain. The three-dimensional structure of human renin is known in detail from X-ray diffraction studies. That is, we know the location of all the atoms in space with precision. Let's see what it looks like.

Biochemists and structural biologists use several different means to illustrate the three-dimensional structures of proteins, each designed to reveal specific aspects of the structure. To begin with, have a look at Figure 3.5, in which I provide the skeleton of the renin molecule.[1] You should be able to find the beginning of the amino acid chain in the upper left of the figure and the end of the chain in the lower center. Each change of direction in the skeleton signifies a new amino acid along the chain. The heavier lines reflect parts of the molecule that are near the viewer whereas the lighter lines reflect those that are more distant from the viewer.

Several things are clear from this representation of the renin molecule. First, the course of the amino acid chain in space is very complex. It wanders from left to right and from top to bottom several times between beginning and end. If you are careful, you should be able to follow the course of the chain throughout its journey. Second, the chain wraps up in a rather compact way. The molecule is not exactly spherical, but it comes reasonably close. Third, parts of the renin molecule that are distant in terms of amino acid sequence are close together in the three-dimensional structure. For example, amino acid 57 along the chain may lie right next to amino acid number 214 in the chain and so on. Last, note that there is a cleft visible at the top of the molecule. This is where the chemistry takes place that renin promotes.

There is a lot missing from the representation in Figure 3.5. We cannot see any structural details, none of the atoms are visible, and it is impossible to know how tightly the structure is wound. Figure 3.6 provides some additional structural information. Here, too, the central feature is the course of the amino acid chain as it winds its way through space. You can still find the beginning of the chain at the upper left, and the end of the chain at the lower center. The cleft at the top of the molecule is still visible. What is new are the helical and sheet structures revealed with thick lines. There are just three helical segments and they all occur in the mid to upper right part of the structure. The sheet structures are shown with arrows and are largely in the left-hand part of the molecule. Note that the sequences intervening between helices and sheets are not organized into specific structures. At the same time, their positions in space are defined.

Figure 3.7 provides a third view of the renin molecule. This is a line representation of the structure in which chemical bonds linking individual atoms are shown as lines. (Bonds to hydrogen atoms are not shown to avoid complicating the structure further.) So we have moved from representations based on individual amino acids along a chain (Figure 3.5) to one revealing higher order structures (Figure 3.6) to one showing atoms as intersecting lines (Figure 3.7). This last view of renin emphasizes the complexity of the molecule and its compactness. Note that the cleft at the top of the molecule remains visible, but it is no longer possible to identify the beginning and end of the amino acid chain.

Last, Figure 3.8 provides a space-filling model of renin. The individual atoms (except hydrogen atoms) are shown as spheres. Carbon atoms are shown in green, oxygen atoms in red, nitrogen atoms in blue, and sulfur atoms (there are not many) in gold. This representation shows just how tightly folded the renin molecule is. There is very little free space between atoms in the molecular interior. It is wall-to-wall atoms in every direction.

Hopefully, the content of Figures 3.5 through 3.8 gives you some sense of the nature of proteins. Human renin is typical of the simplest of proteins—those composed of a single amino acid chain. These proteins wrap up tightly and have various amounts of helical, sheet, and loop structures. Related proteins differ from renin in the content of helical and sheet structures. They also differ from renin in that they contain nonamino acid structures. Perhaps the most common of these are proteins that contain one or more metal atoms. For example, ACE contains a single atom of the metal zinc (Zn). (We examine ACE in chapter 7).

Some Proteins Have More Than One Amino Acid Chain: Quaternary Structure

Still more complex are those proteins with more than one amino acid chain. The individual chains are held together at their interfaces by forces between them. In reality, proteins with more than one chain are the most common in living systems. Hemoglobin, the oxygen-carrying pigment in human blood, provides the iconic example.

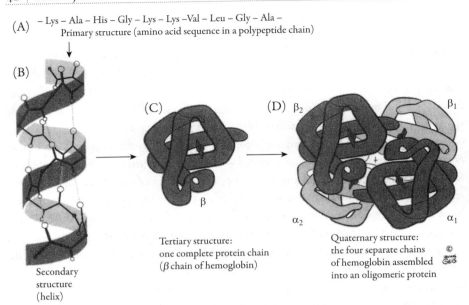

(A) – Lys – Ala – His – Gly – Lys – Lys –Val – Leu – Gly – Ala –
Primary structure (amino acid sequence in a polypeptide chain)

(B)

(C)

(D) β₂   β₁

β

α₂   α₁

Secondary
structure
(helix)

Tertiary structure:
one complete protein chain
(β chain of hemoglobin)

Quaternary structure:
the four separate chains
of hemoglobin assembled
into an oligomeric protein

FIGURE 3.9 The Hierarchy of Protein Structures. (A) A segment of primary structure. (B) Secondary structure illustrated as a segment of α helix. (C) Tertiary structure in which helices are interspersed with coils. (D) quaternary structure. (Illustration, Irving Geis/Geis Archive Trust. Copyright Howard Hughes Medical Institute. Reproduced with permission.)

Specifically, hemoglobin contains four such chains, two each of two types: α chains and β chains. Each chain in hemoglobin has its own tertiary structure. Each chain may be thought of as a subunit of hemoglobin. The hemoglobin molecule is held together by interactions among the individually folded subunits. The quaternary structure of proteins is defined by the spatial arrangement of these subunits. Hemoglobin is by no means unique in having a quaternary structure. In fact, most proteins consist of subunits and thus have a quaternary structure.

The hierarchy of protein structure is illustrated in Figure 3.9. Modern structural biology continues to provide detailed insights into some of the most complex constructs of nature, and we are better off for having them.

The Primary Structure of a Protein Determines the Three-Dimensional Structure

The biological power of proteins is coded in their three-dimensional structures—how the helices, sheets, and loops are organized in space. This organization controls the biochemical activities of the proteins. The three-dimensional structures are, in turn, coded in the primary structures—the amino acid sequences along the chain. The essential demonstration of this fact comes from an experiment of the following sort, using pure preparations of a single protein.

We noted earlier that, under physiological conditions, proteins are folded in complex ways. Under nonphysiological conditions, they can be unfolded to yield an ensemble of floppy, unstructured chains. During the unfolding process, all the secondary and tertiary structures are lost.

If the physiological conditions are restored, each chain assumes the same three-dimensional structure that it had originally. Because there is no agent present to dictate how the unfolded protein chain will refold, the refolding process must be spontaneous. The protein finds its way back to its most stable structure—one with helices and sheets intact. After the primary structure of the polypeptide chain is determined, the three-dimensional structure and, hence, the biological properties are determined as well.

It follows that a change in the primary structure may result in a change in three-dimensional structure and biological properties. We know this is not necessarily the case. Insulins from different mammalian species have somewhat different amino acid sequences, but the same biological function. Cytochrome c from a great many different species has many changes in primary structure, but the same biological function.

In contrast, we also know of many cases in which a very small change in primary structure has major consequences for biological function. Sickle cell anemia is a prominent example to which I return shortly. So we have a question. How are we to know when a change in primary structure—the replacement of one amino acid by another—is likely to alter the three-dimensional structure and biological properties of a protein? To find the answer, we need to understand more about why a protein folds up in the way that it does.

Some Amino Acid Replacements Are Conservative; Some Are Not

Let's go back to the classification of amino acids. As noted earlier, several of them are greasy, fatlike, in nature. Fats and water do not mix. It follows that these side chains will try to do what they can to get out of contact with water. The obvious way to do this is to hide in the center of the protein structure, where they may enjoy the company of like-minded amino acid side chains and avoid that of water molecules—sort of a molecular ethnic cleansing.

At the other end of the spectrum are a number of amino acids that are electrically charged under physiological conditions, bearing either a positive or negative charge. These amino acids insist on being in contact with water. In addition, there are a number of amino acids that are polar but not charged. These side chains, too, prefer to be exposed on the surface of the protein molecule, in contact with the water environment.

The forces that cause proteins to fold up in the way they do are complex and not fully understood. At the same time, the basic idea just introduced is clearly important. One finds the charged amino acids almost always on the protein surface and the highly greasy ones generally in the interior of the structure. This is all that we need to know to explain a good deal about certain human diseases.

Diseases in people come in many flavors. There are infectious diseases (such as measles, mumps, influenza, acquired immune deficiency syndrome [AIDS]), nutritional deficiency diseases (such as scurvy, beriberi, kwashiorkor), degenerative diseases (such as Alzheimer's disease, osteoporosis), cancer (of the lung, breast, prostate, liver, and so on), and single-gene inherited diseases or molecular diseases. In the last category, an important and instructive example is provided by sickle cell anemia. Let's consider this disease and begin to develop a sense of how we can understand it on the basis of what we now know about proteins.

We begin with the symptoms of sickle cell anemia. As the name implies, victims are frequently anemic—that is, they have a content of hemoglobin in blood less than the normal range, the result of rupture of red blood cells, the carriers of hemoglobin (hemolytic anemia). In addition, those with sickle cell anemia are susceptible to chronic infections, may have an enlarged spleen, and experience intermittent bouts of pain, which can be severe, in the bones and joints. The disease can be debilitating.

In the United States, sickle cell anemia is particularly common among blacks. About 8 percent of this population carries a gene for sickle cell anemia. There is an understandable reason for the high prevalence of sickle cell anemia in this population. Individuals harboring a sickle cell anemia gene have a built-in resistance to malaria (for reasons that are not entirely clear), a disease common in sub-Saharan Africa. Hence, carriers of the sickle cell anemia gene had a selective advantage in this region of the world. The frequency of carriers consequently increased over time through an evolutionary process. These people are the ancestors of much of the current black community in the United States.

Hemoglobin is a complex protein molecule with a special task to act as a carrier of oxygen in humans and many other species. Hemoglobin occurs only in the blood and is confined to the red blood cells. Indeed, red blood cells are pretty much wall-to-wall hemoglobin. It is a red protein and it is responsible for the color of blood. As the blood circulates through the lungs, hemoglobin picks up oxygen from inhaled air. Typically, hemoglobin is about 96 percent saturated with oxygen as it leaves the lungs—that is, it is carrying about as much oxygen as it is capable of carrying. As the oxygenated blood circulates through the tissues, hemoglobin gives up some of its oxygen, required to support the life of the tissues (see page 21, chapter 2). It returns to the lung partially depleted of oxygen, is resupplied with oxygen in the lungs, and continues back to the tissues again to deliver life-sustaining oxygen. Clearly, hemoglobin is a molecule essential for human life.

Victims of sickle cell anemia have a subtle but important error in their hemoglobin molecule. This disease is, consequently, referred to as a *hemoglobinopathy*. To understand the subtlety of the molecular error in sickle cell anemia, consider that the composition of normal hemoglobin is $C_{2954}H_{4516}N_{780}O_{806}S_{12}Fe_4$[2] whereas that of sickle cell hemoglobin is $C_{2954}H_{4512}N_{780}O_{810}S_{12}Fe_4$. The very modest difference in composition reflects a single, small, localized change in hemoglobin gene structure that results in the substitution of a greasy amino acid (known as *valine*) where an electrically charged one (known as *glutamic acid*) should be. For the moment, it is enough to understand that a very small change in a critical molecule can lead to a serious disease that afflicts a substantial number of people throughout the world. It is worth understanding the molecular basis for sickle cell anemia in terms of the chemistry of proteins.

We are now in a position to understand the molecular nature of sickle cell anemia. We can ask what is likely to happen to the three-dimensional structure of a protein if we make a conservative replacement of one amino acid for another in the primary structure. A conservative replacement involves, for example, substitution of one greasy amino acid for another or the replacement of one charged amino acid for another. Intuitively, one would expect that conservative replacements would have rather little effect on three-dimensional protein structure. An amino acid happy to be in the protein interior, hidden from water, would be replaced by another having the same preference. Or, an amino acid demanding to be on the protein surface in contact with water would be replaced by another demanding the same environment. This expectation is borne out in reality. As a general rule, conservative changes in the primary structure of proteins are consistent with maintaining the three-dimensional structures of proteins and their associated biological functions.

On the other hand, we can imagine nonconservative replacements. In this case, a charged amino acid might replace a greasy one or the converse. It is easy to imagine that a nonconservative replacement of one amino acid by another in the amino acid sequence of a protein might have substantial consequences for the three-dimensional protein structure and associated biological activity. After all, a residue that seeks a spot on the protein surface replaces one that was content with a spot in the protein interior or vice versa. Structural reorganization is likely to occur.

This is exactly what happens in sickle cell anemia. As noted earlier, in two of the chains of hemoglobin, a unique position in the primary structure occupied by a charged amino acid is replaced by a greasy amino acid. The consequence of this nonconservative change is that the deoxygenated form of the mutant hemoglobin molecule tends to aggregate within the red blood cell. This compromises its ability to serve as an oxygen carrier to tissues. Sickle cell anemia is the result. This is a wonderful example of how we can come to understand the molecular basis of an important human disease on the foundation of a few basic ideas about chemistry.

That is enough for the basics of protein structure. Now it is time to turn our attention to perhaps the most vital of protein functions—catalysis, which is the province of a subset of proteins known as *enzymes*, of which renin is an example.

4 Proteins Perform Multiple Functions
ENZYMES, RECEPTORS, ION CHANNEL PROTEINS

AS EMPHASIZED IN the preceding chapter, there are an unimaginably large number of possible protein structures based on the sequence of amino acids along the amino acid chain. Through the process of evolution, nature has chosen a minute fraction of them to create proteins that provide for the necessities of life. Among all the functions that proteins serve in living organisms, I focus on the three that relate most directly to my tales of drug discovery: catalysis, information transfer, and control of the intracellular milieu. These functions are served by, respectively, enzymes, receptors, and ion channel proteins. I spend most of the time discussing enzymes, because most of the stories in the later chapters focus on enzymes. There are occasions when I refer to receptors and ion channels, as well, but because enzymes are the stars of the stories, let's start there.

Life Depends on a Large Number of Chemical Reactions

Chemical reactions are processes during which one or more molecules are converted into different ones. Chemical reactions involve breaking and forming of the chemical bonds that hold atoms together in molecules. Certain chemical bonds in the starting molecules (the reactants) are broken, followed by the formation of new ones, leading to the end products. All the atoms in the reactants are found in the products; they are just rearranged.

A simple example is provided by diamond and graphite. Diamond is brilliant and the hardest natural substance known; graphite is black and very soft. Yet, both diamond and graphite are composed entirely of carbon atoms (see Figure 2.5). The carbon atoms are linked differently by the chemical bonds holding them together, yielding substances with very different properties. It may surprise you to know that graphite is actually very slightly more stable than diamond. So if we wait long enough, the chemical reaction

Diamond → Graphite

might be expected to occur. However, do not search for evidence of black dots in your wedding diamond. This may be the slowest chemical reaction of all and may take longer than the age of the universe to get anywhere. Your diamonds are safe.

It is obvious that a lot of chemistry happens throughout the course of sustaining life. We ingest complex molecules in our diet but excrete simple ones, including carbon dioxide, water, and urea. All the atoms we take in when we eat and breathe are present in the end products of digestion, but the molecules are very different. The chemical bonds holding the atoms together in molecules have been rearranged.

The totality of the chemical reactions required to support the life of an organism is termed the *metabolism* of that organism, which we discussed in chapter 2. Metabolism is organized into pathways that begin at some defined point and terminate at another one, with something useful happening along the way. You might wish to have a look back at Figure 2.8 to recall the complexity of metabolism.

There is a critical issue here. Left unaided, the chemical reactions required for life are much too slow to sustain life. Let's consider the digestion of proteins, an important constituent in our diet. In an environment similar to that of our digestive system, several tens of thousands of years would be required to digest half the protein content of a typical meal. Clearly, this will not do. Without a doubt, we require some agents, termed *catalysts*, to speed up the chemical reactions of metabolism without themselves undergoing permanent chemical change.

The catalysts of life are the subset of proteins known as *enzymes*, which are enormously potent catalysts. To return to the digestion of dietary proteins, the stomach secretes one enzyme, pepsin, and the pancreas secretes several more into the gastrointestinal tract (commonly known as the *gut*) that catalyze the digestion of other proteins.[1] In the presence of these enzymes, dietary proteins are digested completely and reduced to their basic constituents, the amino acids, in a matter of a few hours, not tens of thousands of years. Enzymes are potent, specific, and subject to regulation.

To understand the power of enzymes a bit more, let's consider a model chemical reaction. When speaking about enzymes, the molecule acted on in a chemical reaction— the reactant—is usually termed the *substrate* (S). The molecule that is the outcome of a chemical reaction is known as the *product* (P). We can write our model reaction simply as S → P, where some substrate molecule is converted into some product molecule.

The point to note here is the difference in the rate of this chemical reaction in the absence and presence of an enzyme catalyst. If we make the necessary measurements, we find that the rate of the reaction in the presence of the enzyme (E) is greater, perhaps enormously greater, than in its absence. At the same time, the enzyme emerges from the reaction unaltered and ready to facilitate the reaction again and again by acting on a succession of substrate molecules. To illustrate, we can write the reaction in the presence of an enzyme as S + E → P + E, showing that the enzyme, E, emerges unchanged as the substrate, S, is converted to product, P. As is emphasized in the next section, the enzyme catalyst recognizes specifically its substrate, S, and is selective in terms of promoting the conversion of S to its product, P.

ENZYMES ARE PROTEINS

The phenomenon of enzyme catalysis in living organisms was known a long time before anyone really knew what an enzyme was in chemical terms. As noted earlier, enzymes are exceptionally efficient catalysts. So efficient, in fact, that chemists working during the early 20th century could observe the phenomenon of catalysis under conditions in which the analytical techniques available at the time could not detect any protein. Proteins were there, but they were not detectable.

The protein nature of enzymes was established through the seminal work of James Sumner. In 1926, Sumner succeeded in isolating the enzyme urease in crystalline form from jack bean meal. This was the first time in history that an enzyme had been obtained in crystalline, although not completely pure, form. Subsequently, Sumner established that the crystalline enzyme was a protein. Urease is an enzyme that degrades one of the human end products of nitrogen metabolism, urea, to ammonia and carbon dioxide (Figure 4.1).

Like many classic contributions to science, this spectacular advance was met initially with criticism and even derision. However, Sumner was not deterred. He took his show on the road and, through demonstration, convinced many scientists of the protein nature of enzymes. Subsequently, a number of enzymes were crystallized during the 1930s by John Northrup and Moses Kunitz, and were also shown to be proteins. These preparations were essentially pure. Northrup and Kunitz demonstrated that enzyme

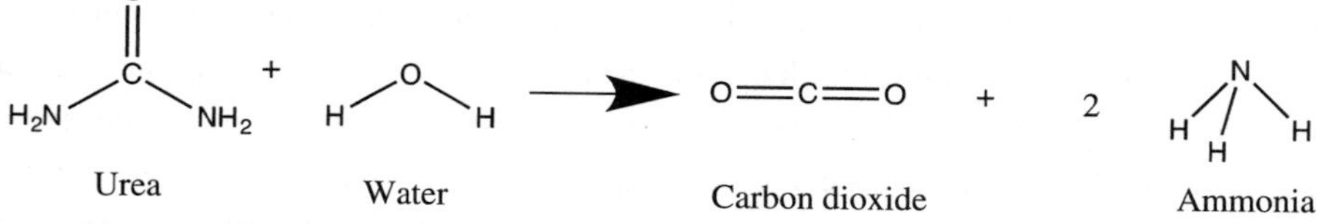

FIGURE 4.1 The Chemical Reaction Catalyzed by the Enzyme Urease. This enzyme catalyzes the conversion of urea, a human end product of nitrogen metabolism, and water to form carbon dioxide and ammonia.

activity paralleled the amount of protein present, laying the issue to rest. Sumner and Northrup shared the 1946 Nobel Prize in Chemistry.

ENZYMES ARE POTENT, SPECIFIC, AND SUBJECT TO REGULATION

Enzymes are remarkable catalysts for three reasons: potency, specificity, and susceptibility to regulation. Earlier, I provided one example of the potency of enzymes as catalysts: the rapid rate of digestion of proteins in the human gut after a meal. There are other ways to appreciate the catalytic potency of enzymes. A second way to understand the point is to measure accurately the ratios between rates of enzyme-catalyzed reactions and the corresponding reactions under the same conditions but without the enzyme present—a refinement of the qualitative argument made previously. These ratios are frequently not easy to obtain because the rates of the reactions in the absence of enzyme may be so slow as to make them exceedingly difficult to measure. Nonetheless, a number of these ratios are known and they typically vary between about 10^3 (1000) and 10^{15} (1 quadrillion or 1,000,000,000,000,000), a truly enormous value. To help understand just how large 10^{15} is, consider a chemical reaction begun at the time of the creation of our solar system, the progress of which would be barely detectable today. That same reaction would be nearly complete in 1 minute if catalyzed 10^{15}-fold.

Enzymes (E) work by forming a complex (E•S) with their substrate (S):

$$E + S \rightarrow E•S \rightarrow E + P$$

in which P denotes the reaction product. Formation of the E•S complex involves the addition of the substrate molecule to a specific site on the enzyme. This site is known as the *active site*. The point is that E•S is very much more reactive than S itself, which is another way of saying that enzymes are highly effective catalysts.

In Figure 4.2, I provide a molecular model example of a complex between an enzyme and a small molecule—in this case, an inhibitor, I. Inhibitors are generally small molecules that have the ability to combine with an enzyme at some specific site and compromise the capacity of the enzyme to carry out its catalytic work. In many cases, inhibitors occupy the same site on the enzyme as the substrate; if the inhibitor is there, the substrate cannot be and no catalysis can occur. Looking ahead, many drugs are enzyme inhibitors, including the drugs discussed in six of the seven tales of drug discovery that follow.

The complex of an inhibitor with an enzyme is denoted as E•I. The E•I complex is a model for an E•S complex. This molecular model shows the enzyme human immunodeficiency virus 1 (HIV-1) protease complexed with a specific inhibitor, known as *amprenavir*. Note that amprenavir occupies a specific site that is partially exposed on the enzyme surface. Amprenavir is used in AIDS therapy. Note that the small molecule amprenavir contacts only a modest fraction of the surface of the enzyme. The same is true for the

FIGURE 4.2 A Molecular Model of the Structure of an Enzyme Inhibitor Complex. This is a line drawing of the amino acid backbone of human immunodeficiency virus 1 (HIV-1) protease (narrower lines) complexed with the inhibitor amprenavir, an AIDS drug, nestled into its active site. Amprenavir is shown in the center of the image in heavier lines. (Source: Shen, C. H., Y. F. Yang, A. Y. Kovalevsky, R. W. Harrison, and I. T. Weber. 2010. Amprenavir complexes with HIV-1 protease and its drug-resistant mutants altering hydrophobic clusters. *FEBS J* 277: 3699–714.)

normal substrate of the enzyme. This site is known as the *active site* of the enzyme and is where the chemistry occurs. Recall that the cleft in the surface of the enzyme renin is its active site (see Figures 3.5–3.8).

In addition, we have the issue of enzyme specificity. The formation of E•S may be thought of by analogy of a key fitting a lock. Here, the lock is the enzyme and the substrate is the key. Just as a key is specific for a lock, so are enzymes specific for their substrates. They are specific for what they bind and they are specific for the chemistry that follows. The lock–key analogy leaves something to be desired, although it provides insight into enzyme specificity. To be more accurate, let's recognize that molecules, unlike locks and keys, are dynamic objects. Chemical bonds vibrate and bend, perhaps a few billion times a second. This causes molecules to do a rapid dance; they rotate, flex, expand, and contract. This dance can be visualized in a process known as *molecular dynamics*. So we have a flexible key (the substrate) fitting a flexible lock (the enzyme).

The formation of an E•S complex is one example of an enormously important phenomenon: *molecular recognition*. The enzyme is said to recognize its substrate just as a lock may be said to recognize its key. The phenomenon of molecular recognition underlies the structure of proteins and nucleic acids; catalysis; the immune response; the senses of sight, taste, and smell; hormone action; the mechanism of action of drugs; and so on. It is one of the keys underlying phenomena of life.

As a general rule, each enzyme catalyzes only one type of reaction and, even then, with a very limited number of substrates. In other words, enzymes are specific for both the type of reaction and the structure of the substrate. There are not many keys that fit these locks and, of those that fit, few open them. Only those that are able to form E•S complexes have the potential to go on to form products. Enzymes are very particular about what they do and with whom they do it.

. . . . Ser-Ser-Asn-Tyr-Cys-Asn-Gln-Met-Met-Lys↓Ser-Arg↓Asn-Leu-Thr- . . .

FIGURE 4.3 An Example of the Specificity of the Enzyme Trypsin. This figure shows the sequence of amino acids along some fragment of the protein ribonuclease A. Each amino acid is represented by a three-letter code. The arrows indicate the sites at which trypsin catalyzes the cleavage of this protein fragment. Of the 14 peptide bonds linking two amino acids, only two are susceptible to the action of trypsin.

To get the idea, let's consider some examples. Earlier I mentioned the digestion of dietary proteins as an example of enzyme catalysis. One of the enzymes secreted by the pancreas that participates in protein digestion is known as *trypsin*. Trypsin will catalyze the cleavage of those chemical bonds linking amino acids only adjacent to lysine (Lys) or arginine (Arg). These are two amino acids that bear a positive charge under physiological conditions. So of the 20 amino acids that occur commonly in proteins, trypsin ignores the bonds between amino acids adjacent to 18 of them. To illustrate, Figure 4.3 is a fragment of the amino acid sequence of ribonuclease A, the enzyme that we met back in chapter 1. I have used the three-letter abbreviations for each amino acid. The arrows show the sites at which trypsin will catalyze the cleavage of the chemical bonds linking two amino acids. All the other bonds linking two amino acids are untouched by trypsin.

Typically, a distinct specific enzyme catalyzes each chemical reaction in the metabolism of an organism. This specificity is required to regulate metabolism properly. One needs to be able to control independently the rates of all, or almost all, metabolic reactions, enzyme by enzyme.

Enzymes are susceptible to a variety of control mechanisms. Some of them control the rates of their synthesis or degradation, and so control the amount of enzyme present. Others control, through activation or inhibition, the activity of the enzyme that is present. The first mechanism works on a timescale of hours to days. The second works on a timescale of seconds.

ENZYME INHIBITORS ARE IMPORTANT CONTROL AGENTS

As I noted previously, there is a class of molecules—some large and many small—termed *enzyme inhibitors*. These molecules bind to enzymes, generally quite specifically, and prevent them from carrying out their catalytic function (see Figure 4.2). These inhibitors are keys that fit the lock but do not open it or alter the shape of the lock so that the key will not fit. This is another example of molecular recognition. In the simplest cases, the inhibitor of an enzyme is related structurally to the normal physiological substrate for the enzyme. The inhibitor looks enough like the normal substrate to bind to the enzyme at the site where the substrate normally binds, but it is sufficiently different so that no reaction occurs. It follows that the enzyme is captured in the form of an enzyme–inhibitor complex, E•I, where I denotes the inhibitor. The point is that E•I cannot make products. The enzyme has been rendered nonfunctional as long as I is bound to it. If I is bound to the enzyme, S cannot be bound at the same time. If S is not bound, no catalysis can take place.

Many of these enzyme inhibitors are important as therapeutic agents. Enzyme inhibitors serve as antibiotics, lower our plasma cholesterol and blood pressure, relieve pain and inflammation, help heal our ulcers, reduce our fevers, and treat cancer, among many other uses. I provide an incomplete but persuasive collection of enzyme inhibitors and their therapeutic roles in Table 4.1.

Most of the remainder of this book tells tales of the discovery of several of these inhibitors. Let's conclude this section with one last look at the enzyme renin. The three-dimensional structure of renin was provided in several formats in Figures 3.5 to 3.8. Recall that there is a cleft in the renin molecule that shows up in the upper part of these figures. That cleft includes the active site of renin, where the substrate binds and where the chemistry takes place.

The substrate of renin is a protein known by the unwieldy name *angiotensinogen*. A space-filling model of this protein is provided in Figure 4.4. Note that here, too, the protein is quite compact, wrapped up tightly with little free space on the inside.

So the substrate of renin is also a large molecule, another protein. The key point here is that the physical interaction between renin and its substrate involves only a small fraction of the renin surface and a small fraction of the angiotensinogen surface. The site of the interaction is at the cleft of the renin structure, revealed in Figures 3.5 through 3.8. Recall that this site is said to be the *active site* of renin and it is where the chemistry involved in catalysis takes place. The task of renin is to split off a small chunk of the angiotensinogen molecule to yield a product known as *angiotensin I*. This molecule has an important role to play in the discovery of an important class of drugs for high blood pressure and heart failure known as *ACE inhibitors*, which are the subject of chapter 7.

There is one small-molecule inhibitor of renin that is approved for treatment of high blood pressure—aliskerin. This molecule binds to renin at the active site of the enzyme, occupying some of the space where angiotensinogen would normally sit. But, with aliskerin in place, there is no place for angiotensinogen to bind and so no chemistry can take place. Aliskerin inhibits the normal enzyme-catalyzed reaction.

An intimate view of aliskerin bound to the active site of renin is provided in Figure 4.5. The surface contours of the atoms of the enzyme are shown. The inhibitor aliskerin is shown as a stick figure; there is an atom at each of the bends in the structure. This figure shows how the small molecule fits neatly into the active site of the enzyme. The shapes are complementary. The key fits the lock.

Receptors Are Essential for Chemical Signaling in Cells

Insulin is a protein hormone. It is secreted into the bloodstream by specialized cells of the pancreas in response to a dietary carbohydrate (sugar) load. The insulin travels to fat and muscle cells that respond to the insulin by taking up sugar from the blood. Blood

TABLE 4.1

A Representative Collection of Enzyme Inhibitors and Their Uses in Clinical Medicine

Use	Enzyme inhibitor
Antibacterial	β-Lactam antibiotics: penicillins, cephalosporins, monobactams
	Ciprofloxacin (Cipro), norfloxacin (Noroxin), moxifloxacin (Avelox), . . .
	Sulfa drugs
Antifungal	Ketoconazole (Nizoral), fluconazole (Diflucan), itraconazole (Sporanox), voriconazole (Vfend), . . .
	Caspofungin (Cansidas), micafungin (Mycamine), anidulafungin (Eraxis)
	Terbinafine (Lamisil), amorolfine (Loceryl), butenafine (Mentax)
Antiviral	Herpes: Acyclovir (Zovirax), penciclovir (Denavir), famciclovir (Famvir), . . .
	Flu: Oseltamivir (Tamiflu), zanamivir (Relenza)
	AIDS: Abacavir (Ziagen), lamivudine (Epivir), . . .
	AIDS: Atazanavir (Reyataz), darunavir (Prezista), nelfinavir (Viracept), . . .
	Hepatitis C: Telaprevir (Incivek), boceprevir (Victrelis)
Anticancer	5-Fluorouracil (Adrucil), capecitabine (Xeloda)
	Imatinib (Gleevec), dasatinib (Sprycel), nilotinib (Tasigna)
	Methotrexate
	Rapamycin (Sirolimus), temsirolimus (Torisel)
	Gefitinib (Iressa), erlotinib (Tarceva), lapatinib (Tykerb), letrozole (Femara), anastrozole (Arimidex)
Pain, inflammation	Naproxen (Aleve), ibuprofen (Advil), aspirin
	Acetaminophen (Tylenol)
Type 2 diabetes mellitus	Sitagliptin (Januvia), vildagliptin (Galvus), saxagliptin (Onglyza)
Central nervous system diseases	Alzheimer's disease: Donepezil, galantamine, rivastigmine
	Depression: Isocarboxazid, phenelzine, selegiline
	Parkinsonism: Carbidopa, benserazide
Cardiovascular diseases	Hypertension, heart failure, kidney protection: Captopril (Capoten), enalapril (Vasotec, Renitec), lisinopril (Prinivil, Zestril)
	Hypertension: Aliskerin (Tekturna)
	High cholesterol: Lovastatin (Mevacor), simvastatin (Zocor), atorvastatin (Lipitor), rosuvastatin (Crestor)
	Arrhythmias, congestive heart failure: Digoxin
	Prevention of blood clots, stroke: Dabigatran (Pradaxa)
	Prevention of blood clots, stroke: Rivaroxaban (Xarelto)

TABLE 4.1 (*continued*)

Use	Enzyme inhibitor
Erectile dysfunction	Sildenafil (Viagra), tadalafil (Cialis), vardenafil (Levitra), avanafil (Stendra)
Benign prostatic hyperplasia	Finasteride (Proscar), dutasteride (Avodart)
Alcoholism	Fomepizole (Antizol)

This table is by no means inclusive of all enzyme inhibitors approved for use in clinical medicine, but it does serve to establish the tremendous utility of drugs in this class.

sugar taken up by these tissues can be used to generate energy or can be stored in the form of glycogen, depending on the needs of the organism.

This sequence of events raises two questions. How does the pancreas know that carbohydrate has been ingested? How do the fat and muscle cells know that insulin has arrived? (Much more information about these questions is found in chapter 12.) There is a general point here. Cells do not live in isolation. Human beings are multicellular organisms. Our bodies contain an enormous number of cells (perhaps 100 trillion, not counting the bacteria that inhabit our guts and elsewhere: 10 times more cells). The number of different cell types depends on how you count them, but there are at least 200. There are several types of brain cells that are distinct structurally and functionally, and the same is true for the liver, muscle, kidney, and so on. Sustaining life requires that these cells work together in a highly cooperative and coordinated way. It is similar to an orchestra playing a symphony. There are many different instruments and each has a specific role in the orchestra. When all the instruments play together in a coordinated way, we get music; if not we get noise.

The requirement that cells act in a coordinated way implies that cells be linked through a complex communication network. There are a number of ways that cells communicate with each other. One of these is by means of extracellular signaling molecules. Insulin is one example of an extracellular signaling molecule.

Communication requires a signal and a signal receiver. Signal receivers are termed *receptors*. The signal receptor for insulin is the insulin receptor, a protein localized in the membrane that surrounds, for example, fat and muscle cells. When insulin arrives at the cell surface, it can bind to the insulin receptor and communication happens. When insulin binds to its receptor, the receptor is activated and it elicits changes in the cell metabolism. Sugar is taken up from the blood and is either used or stored.

Receptors have at least two things in common with enzymes. In fact, many receptors are enzymes-in-waiting and become active catalytically through binding of a signaling molecule. The insulin receptor falls into this class. In Figure 4.6, the essentials of the action of the insulin receptor are provided. Note the following. First, the insulin receptor spans the membrane

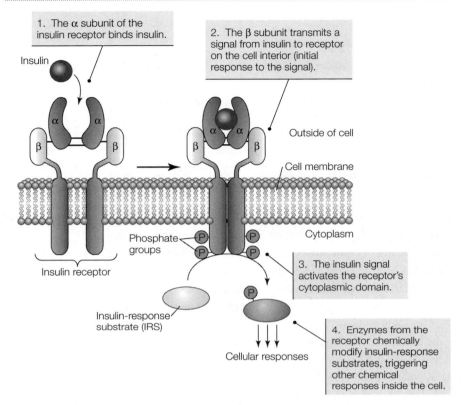

1. The α subunit of the insulin receptor binds insulin.

2. The β subunit transmits a signal from insulin to receptor on the cell interior (initial response to the signal).

Insulin

α   α

β   β

α   α

β   β

Outside of cell

Cell membrane

Cytoplasm

Phosphate groups

Insulin receptor

3. The insulin signal activates the receptor's cytoplasmic domain.

Insulin-response substrate (IRS)

Cellular responses

4. Enzymes from the receptor chemically modify insulin-response substrates, triggering other chemical responses inside the cell.

FIGURE 4.6 The Mechanism of Action of Insulin. This illustration summarizes the steps in the action of the hormone insulin, acting through the insulin receptor. The net effect of a molecule of insulin from the circulation binding to the insulin receptor outside of the cell is a number of chemical reactions that occur within the cell. Thus, the insulin receptor acts as an information transfer device.

surrounding the cell. Second, it binds insulin that arrives at the outside of the cell. Insulin is the key that fits the insulin receptor lock. Third, the receptor transmits the information that an insulin molecule has bound across the membrane to the cell interior. Fourth, activation of the receptor follows and it becomes active enzymatically. Last, numerous chemical changes occur within the cell, leading to the consequences of insulin action.

Both enzymes and receptors are protein in nature and both are highly specific—enzymes for their substrates and receptors for their signaling molecules, known as *ligands*. The insulin receptor is specific for insulin, the growth hormone receptor is specific for growth hormone, the testosterone receptor is specific for male sex hormones, and so on.

In summary, receptors are proteins, membrane bound or intracellular, that act as specific receivers for chemical messages and initiate metabolic changes in cells and organs. Specific agonists (molecules that activate the receptor) and antagonists (molecules that inhibit receptor activity) at receptors comprise several classes of drugs useful in clinical

TABLE 4.2

A Representative Sample of Useful Drugs That Target Receptors

Drugs that are antagonists at the histamine H1 receptor. These drugs are used to relieve allergic symptoms such a hay fever.

Loratadine (Claritin)

Chlorpheniramine (multiple trade names)

Diphenhydramine (multiple trade names)

Drugs that are antagonists at the histamine H2 receptor. These drugs are used to treat esophageal reflux disease and peptic ulcers (in combination with antibiotics).

Ranitidine (Zantac)

Famotidine (Pepcid)

Cimetidine (Tagamet)

Nizatidine (Axid)

Drugs that are agonists at the adrenergic beta receptor. These drugs are used to treat asthma and other bronchoconstrictive conditions.

Albuterol (Ventolin)

Formoterol (Perforomist)

Formoterol plus budesonide (Symbicort)

Drugs that are antagonists at the adrenergic beta receptor: beta blockers. These drugs are used for high blood pressure, heart failure, angina, glaucoma, and stage fright.

Propranolol (Inderal)

Atenolol (Tenormin)

Metoprolol (Lopressor, Toprol)

This list is far from comprehensive, but should serve to establish the clinical utility of molecules that target receptors. Note that these may be agonists (molecules that bind and elicit a biological response) or antagonists (molecules that bind and prevent a biological response).

medicine. Although none of key actors in the stories that follow target a receptor, there is mention of a few that do. To emphasize the point, I provide in Table 4.2 several examples of clinically useful drugs that target receptors.

Ion Channels Regulate the Intracellular Milieu

Biological membranes, whether they enclose cells or organelles, such as mitochondria, within cells, are basically barriers. Those that enclose cells, termed *plasma membranes*, isolate the cell interior from the extracellular environment. This barrier function of membranes is essential to life.

At the same time, there is clearly a need for some sort of communication and material exchange between the cell interior and its external environment. To meet these needs, living organisms have developed a number of specialized, membrane-centered

TABLE 4.3

Examples of Clinically Useful Drugs That Target Ion Channels

Sulfonylureas for diabetes: ATP-dependent potassium channel in pancreatic beta cells
    Glipizide (Glucotrol)
    Glyburide (Diabeta)
Calcium blockers for high blood pressure, migraine, certain arrhythmias:
cardiac calcium ion channels
    Verapamil (Calan)
    Nifedipine (Adalat)
    Flunarizine (Sibelium)
    Diltiazem (Cardizem)
    Amlodipine (Norvasc)
Sodium blockers for arrhythmias: cardiac sodium channel
    Flecainide (Tambocor)
    Mexiletine (Mexitil)
    Propafenone (Rythmol)

The entries in this table reflect a modest fraction of all clinically useful drugs that target ion channels. These examples do establish the utility of this class of proteins as targets for drug discovery. ATP, adenosine triphosphate.

mechanisms. As noted in the previous section, membrane-localized receptors provide one means for message exchange between cells and the external world. Ion channels provide another.

Ions are atoms or small collections of atoms that bear an electric charge, positive or negative. Examples of ions that are critical for life include the sodium ion Na^+, the potassium ion K^+, the chloride ion Cl^-, the magnesium ion Mg^{2+}, the calcium ion Ca^{2+}, and the iron ions Fe^{2+} and Fe^{3+}.

Ion channels are complex proteins that span the plasma membrane. Plasma membranes themselves are composed largely of hydrophobic molecules and are nearly impermeable to ions. This impermeability provides a means of controlling access of ions to the cell interior.

Ion channels are ion-specific protein gates that open and close in response to suitable signals. This is known as *gating*. Think about a gate in a picket fence. The gate needs to be open for people to pass through. Ion channels work the same way. In the resting state, they are closed and ions cannot pass.

These gates may open in response to a voltage change across the membrane: voltage-gated ion channels. The sodium and potassium ion channels of nerve cells fall into this category. These gates may also open in response to the binding of some small molecule or ion, known generically as ligands: ligand-gated ion channels. For example, ion channels that open in response to a neurotransmitter, such as dopamine or serotonin, are ligand-gated ion channels.

Molecules that target ion channels have proved to be of clinical use, although they have been less important as targets than enzymes or receptors. We shall encounter one example in chapter 10 on avermectins. Table 4.3 presents a few examples of drugs that target ion channels.

We now have almost all the background required to understand drug discovery. I provide what little is lacking in the next chapter: the process of drug discovery and development. Then, it is on to the adventure stories of how drugs actually find their way into clinical practice.

5 Drug Discovery and Development
THE ROAD FROM AN IDEA TO PROMOTING HUMAN HEALTH

WE ARE ALMOST where we need to be to grasp the tales of drug discovery that make up the final seven chapters of this book. The three previous chapters have laid the necessary scientific foundation. Here is my take on what you still need to know to understand drug discovery and development: the process of getting from an idea to a product that meets a medical need—from the laboratory to the bedside.

We begin with a look at the process from 35,000 feet. Realize one thing at the outset: there is more than one way of getting from an idea to approval of a new drug for use in human medicine. The process described in this chapter captures the essential features of getting this done. However, each drug discovery effort poses specific problems, and getting around them may have an effect on the actual process followed. Nonetheless, what is described here is well worth understanding.

Doing Drug Discovery and Development Is a Bit Like Running a Really Challenging Maze

Imagine running a maze that may have more than two dimensions, a huge number of entry points, and a very small number of exits or perhaps no exit at all, depending on your entry point. Have a look at Figure 5.1 to get the idea. You can see the external features but have no clue what awaits you inside. You can wander in this maze for a long time without getting very far. This pretty well symbolizes what happens during a lot of drug discovery projects.

In target-based drug discovery, described in the next section, each entry point in the maze corresponds to a molecule chosen as a starting place to begin work. Most of these entry points are dead ends. No matter what you do or how hard you try, the only exit from the maze is the place you entered—nothing gained. The journey may be long, there may be encouraging signs along the way, but at the end of the day, you are back where you started. You need to find a new entrance—that is, find a better molecule to begin with and hope this one works out better than the last one you tried. If you do this long enough, you may find your way to an exit, which corresponds to a molecule that enters

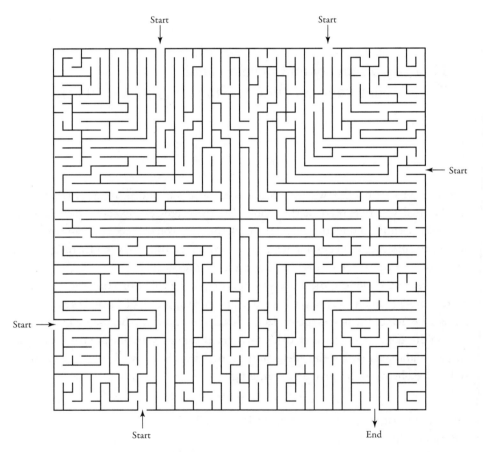

FIGURE 5.1 The Maze as a Metaphor for the Drug Discovery Process. Here is an example of a complex maze intended to convey some sense of the difficulty in the drug discovery process. This maze has five entrances but just one exit. Four of the entrances lead to dead ends but you do not know which four. Think of each entrance as one potential drug discovery route, perhaps starting with an interesting molecule. Think of the exit as a product molecule. In four of the five cases, it does not matter which path you follow; you are doomed to failure. In the fifth case, success is not ensured; you still have to find your way through the maze.

the pharmaceutical marketplace to address a medical need. There may be a few of these for each medical need. Note that if the target is chosen improperly, there is no exit from the maze except the way you came in.

Spending your professional life running a maze of this degree of difficulty may seem like a poor choice of a way to make a living. Nevertheless, it is what scientists in the pharmaceutical industry do, and they do it with a surprising degree of optimism and enthusiasm. The reward is to have had some significant role in bringing into medical practice a drug that meets an important medical need. This goal provides motivation. Wouldn't you like to retire at the end of your working life knowing that your efforts helped to create a better drug for cancer or Alzheimer's disease?

Drug Discovery May Be Target Based or Phenotypic Screen Based

There are two general ways to approach drug discovery. The first is *target-based* drug discovery. In this case, the target is a defined molecule, generally a protein. Typical targets are enzymes, receptors, and ion channels (chapter 4). The point is to find a small molecule that has the ability to modulate—activate or inhibit—the activity of the target molecule in a way expected to have a beneficial effect on some human pathology. Among the stories that follow, the discovery of finasteride, ACE inhibitors, the statins, fludalanine, and sitagliptin provide examples of target-based drug discovery. The molecular mechanism of action of a drug is established at the very outset of the discovery process; you know at which enzyme, receptor, or ion channel the drug acts. Most drug discovery projects in the pharmaceutical industry today are target based.

The alternative is *phenotypic screen-based* drug discovery. A phenotypic assay is a laboratory test that determines the effect of test molecules on some observable trait of a cell, tissue, or organism. For example, suppose you wish to search for a molecule that has the potential to lower blood pressure in people with high blood pressure, a threat to good health. You might elect to choose an animal model of high blood pressure—the spontaneously hypertensive rat provides one alternative—and test your candidate molecules in this model. Your assay will consist of measuring the effect of test molecules on the blood pressure in your animal model. Those that lower blood pressure effectively will be classified as *hits* or *actives* and are candidates for further evaluation. You will not know the molecular target (the mechanism of action) of your actives, but you do know they are effective in your model of a human disease. Subsequently, you may elect to try to pinpoint the mechanism of action (not very easy in most cases). Contrast this with the target-based approach to high blood pressure described in chapter 7 in which the enzyme ACE is the target. Among the stories that follow, the discovery of imipenem and the avermectins provide examples of phenotypic screen-based drug discovery. Let's look at some more examples.

Helicobacter pylori infection causes ulcers in the stomach and duodenum, the proximal part of the intestine. One might search for a molecule that kills or prevents replication of this bacterium in a test tube or agar plate assay. This is a phenotypic screen. You would not know the molecular mechanism of action of an active molecule, but you do know that it has the desired activity. You can screen for antifungals or antiparasitics in the same way, using the intact organism. Mice bearing specific human tumors can be used as phenotypic screens in a search for molecules that elicit regression of those tumors. Drug discovery during the 1970s and earlier depended very largely on phenotypic screens. They continue to be used and are enjoying something of a renaissance in the pharmaceutical industry today.

In most cases, target-based and phenotypic drug discovery join in a meaningful way after the earliest stages of the process are complete. Specifically, active molecules discovered in a target-based screen will always be evaluated for activity in cell lines, tissues, or whole animals. Thus, target-based drug discovery, sooner or later, relies on phenotypic assays to take the biological measure of interesting molecules. In turn, the molecular mechanism of action of molecules with provocative activities in phenotypic screens is always of interest. Searches for mechanisms are pursued as the discovery process moves forward. Knowing the mechanism of action usually aids in anticipating safety problems, gaining regulatory approval for marketing, or finding alternative uses for a molecule discovered during a phenotypic screen. In short, you need both types of assays to get where you need to be. In what follows, I focus on the process generally followed for target-based drug discovery. As just mentioned, the only real differences between target-based drug discovery and phenotypic drug discovery occur during the early stages of the process.

There Are Nine Stages in Target-Based Drug Discovery and Development

Like any complex process, that of target-based drug discovery and development can be divided into stages. Different scientists divide up the process a bit differently. However, this makes no real difference, so we proceed. The process illustrated in Figure 5.2 is as good as any and sets the stage for our discussion. Nine stages comprise the overall process. Let's have a look at them one at a time. As we proceed, keep in mind that the process varies some from project to project. One size does not fit all in drug discovery and development.

THE DISCOVERY PROCESS BEGINS THROUGH IDENTIFICATION OF A TARGET

In every drug discovery search, you have to have a place to start, specified as Target ID in Figure 5.2. Starting places for drug discovery are not assigned like post positions in a horse race. You have to find one yourself or else adopt one that has been selected by a competitor. A typical starting place is the choice of some protein; let's say an enzyme is the target. To provide a specific example, consider the enzymes involved in the synthesis

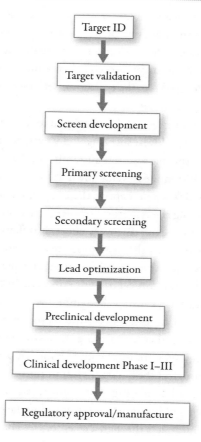

FIGURE 5.2 Overview of the Process of Target-Based Drug Discovery and Development Divided into Nine Stages. The first six stages fall into the category of discovery, and the final three are associated with development. In practice, discovery and development overlap a good bit. The details of this process vary some from project to project and among different pharmaceutical companies.

of the bacterial cell wall, a structure that does not occur in human cells. These enzymes are unique to bacteria and provide good targets in a search for antibiotics. If we can find molecules that inhibit one or more of these enzymes, bacteria will not be able to create their cell walls and will either die or cease to grow. Because the enzymes are unique to bacteria, such inhibitors should not affect human metabolism. The Primaxin and flu-dalanine tales that follow in later chapters provide details.

Since 1913, evidence has accumulated indicating that excess blood cholesterol is a threat to good health. The biosynthetic pathway to cholesterol is known in detail, as are the enzymes that catalyze each of the reactions along the way. It makes sense to think that an inhibitor of one or more of these enzymes might reduce blood cholesterol by inhibiting cholesterol synthesis. The tale of the statins provides the details in chapter 8. For each of the drug discovery stories, the rationale for choice of molecular target or the nature of the phenotypic screen is provided.

Before going any further, there is something that you should understand: most drug discovery projects fail, and there are two basic reasons why. First, the underlying hypothesis of the project is flawed. In other words, target identification is flawed. Molecules that influence the activity of the target molecule do not prove to have the expected consequences. The project is doomed to failure at the outset. There is no way out of the drug discovery maze; many entries, no exit.

Here is an example. Histamine is a molecule known to cause the symptoms of allergic reactions such as hay fever and hives: sneezing, itchy eyes, runny nose, general misery. The amino acid histidine is the sole source of histamine in the human body. The enzyme that converts histidine to histamine is well known. A plausible hypothesis is that an inhibitor of that enzyme, which would deprive the body of its only source of histamine, would be effective in relieving the symptoms of allergic reactions. Although a plausible and attractive hypothesis for a drug discovery effort, it is flawed. Not knowing this in advance, Merck scientists targeted the histamine-creating enzyme in a drug discovery effort years ago. They discovered a great inhibitor of the enzyme that catalyzes histamine synthesis, and they moved it into clinical trials. It proved ineffective. Now we know why. The cells that secrete histamine, causing the symptoms of allergy, release only a small fraction of stored histamine at any one time. So even if you shut down histamine synthesis, for all intents and purposes, there remain large stores of this molecule ready to trigger successive attacks of allergy symptoms. The underlying hypothesis was flawed because we did not understand enough of the biology of histamine storage and release. The result was a great enzyme inhibitor but no drug. Antagonizing the *action* of histamine, rather than inhibiting the *synthesis* of histamine, is effective in relieving the symptoms of, for example, hay fever. Histamine acts through a receptor (see chapter 4) known as the *histamine H1 receptor*. Histamine forms a complex with this receptor just as enzymes form complexes with their substrates. Once the histamine•receptor complex forms, physiological consequences result, just as catalysis results from formation of E•S complexes. If a receptor antagonist prevents histamine from complexing with its receptor, then the action of histamine is prevented, which is what antihistamines do. You can buy them over the counter (Benadryl Allergy, Chlor-Trimeton Allergy, Dimetapp Allergy, Loratadine, Alavert, Sudafed) or with a prescription (Allegra, Clarinex, Zyrtec). The alternative strategies are summarized in Figure 5.3.

Second, efforts to discover a molecule that meets all the requirements for use in human health may fail. The two basic requirements are that the molecule (1) be *effective* for the intended use and (2) be *safe* for the intended use. Finding a molecule that is safe for the intended use is the biggest single barrier to success in drug discovery. Beyond these basic issues, there are several other things that must be true. For example, the drug ordinarily must have a suitably long duration of action. A pill taken once a day is ideal; for some indications, it is essential. No one is going to take a pill every 3 hours for the rest of his or her life. In addition, the drug molecule must be stable. Most drugs have a shelf life of at least 2 years. Third, you must have a commercially viable source of

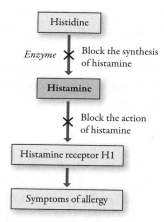

FIGURE 5.3 Alternative Strategies for Modulating the Action of Histamine to Relieve the Symptoms of Allergy. This figure shows two different ways of thinking about target-based drug discovery for the control of allergic symptoms. The molecule responsible for these symptoms is histamine. The amino acid histidine is the only source for histamine; a well-known enzyme catalyzes this conversion. This enzyme is, therefore, a plausible target for drug discovery work. Histamine works through the action of a histamine H1 receptor. Therefore, this receptor is also a plausible target for drug discovery. The former alternative blocks the action of the enzyme that makes histamine; the latter blocks the action of histamine. As described in the text, blocking histamine action proves effective in controlling the symptoms of allergy. Blocking histamine synthesis does not.

your drug. If the molecule is so complex that a billion-dollar manufacturing plant is required to make it, you need to find a simpler molecule.

It follows that success depends on valid target identification in the first place and the ability to discover the right molecule, which brings us to the second step: target validation.

TARGET VALIDATION IS A KEY REQUIREMENT FOR DISCOVERY SUCCESS

There are several ways to attempt to validate a potential target for a drug discovery effort. Let's look at a few of them.

Somebody Else Did It First

Years ago, Squibb (now part of Bristol Myers Squibb) proved that an inhibitor of an enzyme known as ACE is safe and effective in lowering elevated blood pressure in people (see the story in chapter 7). Their ACE inhibitor is captopril, marketed as Capoten.[1] The underlying hypothesis of the Squibb discovery program—that an inhibitor of ACE would be safe and effective for treating high blood pressure—was demonstrated to be correct in clinical trials, which removed all doubt. It also opened the door to competitors who did, indeed, rush in to discover their own ACE inhibitors. Having a confirmed drug discovery hypothesis—a validated target—is highly attractive because so many of them are wrong. Using a known active compound as a starting point for drug design is a validated use of time and resources. One can search for more effective molecules, safer molecules, drugs that offer a more convenient dosage regimen, and so on. It is often the case that the best drug in a given category, and the biggest commercial success, is the second or third one to enter the market. Several other companies did succeed in entering their own ACE inhibitors into clinical practice (chapter 7).

Nature Provides Guidance

Accidents of nature, the result of mutations,[2] may provide important guidance in formulating a drug discovery hypothesis. For example, it was known that men with a defective enzyme known as 5AR2 have a small prostate that does not grow, do not get acne, and do not go bald (undergo male-pattern baldness). A plausible hypothesis based on these observations is that an inhibitor of 5AR2, mimicking the genetic error, would be effective in treating an enlarged prostate in aging men and in preventing acne and male-pattern baldness. This hypothesis, too, is correct and the story is told in chapter 6.

You May Be Able to Construct a Plausible Mimic of What You Are Trying to Achieve

Suppose you have biomedical information that suggests that inhibiting a specific enzyme will treat or prevent a specific disease; this is your hypothesis. There may be one or more ways of improving your confidence in the hypothesis without mounting a full-scale drug discovery effort. For example, you may be able to remove the gene that codes for that enzyme from the genome in an animal model of your disease and evaluate the results. Without the gene, there can be no enzyme that is coded for by that gene. The biology of an animal lacking the gene should reflect the biology that would be elicited by completely inhibiting the enzyme coded for by that gene. You may gain insights into both safety and efficacy of a potential drug through such studies.

Alternatively, you may be able to invoke technologies that prevent the synthesis of an enzyme without deleting the gene for it in an animal model and, again, evaluate the results. Without extending this list further, let me simply state that there are a family of technologies that may be invoked to strengthen or weaken a drug discovery hypothesis.

There Is Strongly Suggestive Evidence from Basic Biomedical Research Supporting Your Hypothesis

There is a remarkable, international basic research effort to understand human biochemistry, physiology, and pathology. Novel insights into human health result. Some suggest ideas for prevention or treatment of human diseases. These glimpses are frequently intriguing and difficult to resist. They are also sometimes wrong. Despite all we know, we do not know enough to predict with accuracy the result of proposed specific drug interventions for human health. Note the case of the histamine-creating enzyme mentioned earlier. This situation will improve as we learn more.

Suffice it to say that pharmaceutical houses go to considerable lengths to validate their identified targets. No one wants to start a discovery project that has no hope of success (although this happens a lot anyway). Throughout the course of a discovery effort, molecules are created that have the potential to validate (or invalidate) the selected target in animal models of the chosen disease. These studies are undertaken routinely. If one is on the wrong road, the sooner this is learned, the better. Traveling a long road leading nowhere is neither fun nor profitable. Target validation is an ongoing process throughout the discovery process.

The Human Genome Project has had a huge impact on drug discovery going forward.[3] Its numerous successor genome-sequencing projects have as well. In its most basic sense, the Human Genome Project has identified every protein-coding gene in humans. There are about 21,000. The number of proteins in the human body is substantially greater than this number because many genes can code for more than one protein.[4]

Through identification of every protein-coding gene in the human genome, we have greatly expanded the number of potential protein targets for drug discovery. In principle, we have in hand all the possible human protein targets for drug discovery, although not all predicted proteins have yet been identified and characterized for their physiological roles.

In the short run, I believe the availability of this cornucopia of molecular targets may have had a negative effect on drug discovery. Too many companies have launched too many discovery efforts on minimally validated targets with predictable results: many failures. In the longer term, there can be no doubt that the knowledge gained from the Human Genome Project and its successors will be a great benefit for human health, including the ability to discover breakthrough drugs that meet critical health needs.

For example, the genomes of many cancer cell lines have now been sequenced. Having the knowledge of the genome of normal cells in hand permits us to identify changes in sequences in cancer cells (mutations) and to associate some of them with the malignant state, which opens two doors. First, knowing which altered proteins are associated with cancer permits us to target those proteins in searches for drugs for cancer. For example, we know that derangement of a protein known as *c-Met* is associated with cancers of the liver, breast, brain, stomach, and kidneys. It is plausible that inhibitors of c-Met will be useful agents in the treatment of these cancers. Several molecules in this class are in clinical trials for cancer.

Second, tumors of different organs may share underlying molecular defects. Note the case of c-Met in the previous paragraph. When molecular defects are shared, a drug that acts to correct or minimize the consequences of that defect may find use in multiple cancers. Efforts are now under way to explore new uses for approved cancer drugs based on genomic data from tumor cell lines.

In addition to the human genome, those for many other living organisms have been identified. The publication of a new genome is almost a weekly occurrence now. These genomes include those for almost all organisms know to be pathogens in people. Thus, we have in hand all the possible molecular targets for the discovery of novel antibacterials, antivirals, antifungals, and antiparasiticides. New and effective drugs in these categories are certain to follow. Genomic studies have created a new and promising day for human health.

SCREEN DEVELOPMENT IS NECESSARY IN THE SEARCH FOR CANDIDATE MOLECULES

We need to have some means to measure what we are trying to achieve. A screen is a laboratory test to determine whether a series of compounds has the desired activity. A screen can be used to look at a few hundred compounds or a few million. For example, the search for molecules that inhibit enzyme activity requires some means of measuring that activity, and are known as *enzyme assays*. Such assays are used in screening collections of molecules for the desired activity. When you have an assay suitable for running through collections of molecules, then you have developed a screen.

There are a great many ways of assaying enzyme activity. Basically, you find some way to measure the disappearance of the substrate or the appearance of the product (or both) as a function of time. Molecules have fingerprints, and chemists have devised a number of ways of detecting those fingerprints. Some of these technologies are quite simple and some are strikingly complex. Here is/a simple example.

Let us suppose that the substrate for an enzyme is colorless but that the product is colored—say, yellow. We could follow the activity of the enzyme by watching the development of yellow color over time. Instruments known as spectrometers do this very accurately. They collect and send the data to computers for analysis and presentation. Typically, we would make measurements of the progress of the enzyme-catalyzed reaction in the absence of an inhibitor and in the presence of increasing concentrations of inhibitor. A typical outcome is shown in Figure 5.4.

Quantitative analysis of data of the sort provided in Figure 5.4 yields a measure of the potency of an inhibitor. A typical measure (although not the best one) is the concentration of inhibitor required to reduce the rate by half—that is, to 50 percent of its uninhibited rate. This measure is known as IC_{50}.[5] Scientists engaged in drug discovery focus on potency as an important property. All things being equal, increasing potency means decreasing the dose required to realize a beneficial effect. Small doses reduce the risk of adverse effects, minimize the costs of drug manufacture, and reduce environmental contamination resulting from the manufacturing process.

A search for an enzyme inhibitor provides one example out of many of the need for an assay to measure a desired activity. The idea is just the same in searches for receptor agonists or antagonists, ion channel blockers, or molecules that kill *Helicobacter*. However, for whatever you search, you need an assay to guide the search.

Having an assay provides the means to search for biologically active molecules. The next question is how one conducts the search.

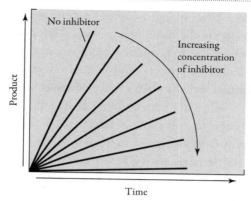

FIGURE 5.4 Schematic Example of an Assay of Enzyme Activity. The appearance of the product of the enzyme reaction is followed as a function of time. As the concentration of inhibitor is increased, the rate of product formation is decreased. Analysis of these data provides a quantitative measure of the potency of the inhibitor.

PRIMARY SCREENING IS FREQUENTLY RAPID

There are a number of more-or-less effective ways of searching for molecules that will exhibit activity in an assay. A common one is to search through huge collections of known molecules, screening in the hope of stumbling across one or more molecules that have the desired activity. Big pharmaceutical companies generally have very large collections of molecules assembled over time. A sample of each compound made in the companies' laboratories is saved in an archive. These archives may contain hundreds of thousands of molecules, which is a valuable resource in drug discovery efforts. In general, a subset of the molecule archive will be screened for activity in a new drug discovery project. Robots usually carry out the mechanics of the assay. Running 100,000 molecules or more through as assay demands automation. In favorable cases, 100,000 compounds may be assayed in a matter of days using very small amounts of each compound, perhaps 1×10^{-5} g.

If fortune smiles, this effort will uncloak a small collection of biologically active molecules of diverse structures: entry points into our complex maze. These molecules generally lack a family of properties (described later) required of a molecule worth getting excited about. They serve as starting points for medicinal chemists who strive to build in the set of properties required for moving ahead toward a clinical trial. In a typical case, chemists will make 1,000 to 2,000 compounds either (1) to discover a molecule attractive enough to move into clinical trials or (2) to give up the search and find a different way to spend time and money.

Compound collections may be accessed in other ways. For example, there are a number of companies that have sought out compound collections from diverse sources and make a business of selling them. Acquiring a compound collection is an attractive way for small or new drug discovery companies to establish a molecule archive for screening.

An alternative is to search through sources of natural products—molecules made by living organisms—with the same hope. Natural products generally come as complex mixtures; a bacterial fermentation broth, for example, may contain hundreds of molecules. Similarly, extracts of plants usually yield complex mixtures of molecules. Fermentation broths or plant extracts are screened for activity just as individual molecules are. When one finds an active, there follows an exhaustive process of purification in an effort to isolate and characterize the molecule or molecules responsible for the activity (see chapters 9 and 10 for examples). Search of natural products is by no means restricted to microbial fermentation broths and plant extracts. Chemists have explored just about every source of natural products imaginable. For example, the antidiabetic drug Byetta (exenatide) is a modification of a protein molecule isolated from the saliva of Gila monsters. About 40 percent of all drugs approved in the United States for use in human medicine are either natural products or are derived from natural products.

Screening compound collections and natural products differs in several ways. Molecules in archives are usually screened one at a time. When you find biological activity, you know what it is. When you screen complex natural products mixtures and find biological activity, all you know is that there is some molecule (or molecules) in the mixture that works (work), but you do not know what it (or they) is (are). As noted earlier, you have to isolate the interesting molecule from all the others and deduce its structure. You have to do a lot of work to find out what you have. One of the disheartening outcomes of natural product chemistry is that you may discover a molecule that is already known. There are a lot of natural product chemists and they have been at it for a lot of years, so a lot of natural products have been characterized. You run the risk of repeating the work that someone else has already done.

But there is a happy side to natural product chemistry. Plants and microorganisms, for example, go to the trouble of making complex molecules for a reason, frequently to protect themselves from predators. So natural products are biased in favor of being biologically active. Biologically active molecules found in screening compound collections need extensive remodeling by chemists over a couple of years before coming up with something interesting (or failing to do so). In contrast, actives isolated from natural sources sometimes need little or no chemical tweaking.

The classic examples of natural product-based drug discovery come from the search of fermentation broths for antibacterials (frequently known simply as *antibiotics*). Penicillin was isolated from the fermentation broth of a mold. It required no chemical tweaking to make it suitable for use in human medicine. However, chemical tweaking of the penicillin structure generated a whole family of semisynthetic penicillins: penicillins G and V, Trimox (amoxicillin), Pen A (ampicillin), Azapen (methicillin), Zosyn (piperacillin), and on and on. Cephalosporins provide a second example. Here, too, modest chemical variations on the structure of a natural product generated a family of highly effective antibiotics: Keflin, Ancef, Cefam,

Zinacef, Ceclor, Mefoxin, and on and on. So there is a trade-off. Natural product chemistry has some issues in terms of isolation and structure determination work that may be offset by finding something really interesting without doing years of medicinal chemistry in the form of chemical modifications and structure optimization.

Like all human activities, searching through molecule archives and sources of natural products can be done well or done poorly. Success depends on the quality of the assay used to search for biologically active molecules; the size, nature, and structural diversity of the collections screened; the ability of scientists to isolate and characterize active molecules from complex mixtures; and good fortune. The final issue should not be overlooked. If your collection of molecules to be screened for activity does not contain one or more actives for the target, then it makes no difference how well or how carefully you work. You cannot succeed. Given this reality, scientists have worked hard to develop methods of inhibitor design that are as free as reasonably possible of the element of good fortune. This is known as *de novo drug design* or *computer-based drug design*—an alternative to screening molecules in the laboratory to identify those that interact in a desirable way with the target.

One sophisticated approach to drug design depends on having a detailed structure of, for example, the active site of a target enzyme. Detailed structure means atom-by-atom resolution. These structures generally result from careful X-ray diffraction studies of the target molecule. The detailed structure of a potential binding site for an inhibitor provides a negative template for inhibitor design. Think of the negative template as a lock; inhibitor design involves creating a key to fit this lock. Refer back to Figure 4.5 for an example.

The detailed structures of proteins and their interactions with small molecules are visualized through molecular graphics techniques (see Figures 3.5–3.8 and Figures 4.2 and 4.5), which are key to success in designing molecules to fit binding sites. These structures permit the chemist to explore visually just how well numerous molecules fit into the binding site. Complex computer algorithms that calculate the energy of binding between the binding site and test molecules aid the process. None of this works perfectly, but it does work well enough in many cases to provide reasonable starting points—novel molecules—for medicinal chemistry work and to aid medicinal chemists as they work to improve their molecules. There are many variations on this theme of inhibitor design.

There is a final way of thinking about molecular design. Once more, consider enzyme inhibitors. We know that enzymes bind the substrate and the product molecules. It follows that other molecules related structurally to substrate and/or product molecules have the potential to bind to the active site as well, which gives chemists a decent start at thinking about what they might wish to do. A particularly interesting variation on this theme derives from recognizing something about the nature of enzyme catalysis, which I describe in the next paragraph.

As the substrate undergoes structural changes on the way to product, it climbs an energy hill. Chemical bonds must be loosened or broken before others are formed. The bond breaking requires energy, a hill to climb. At some point, the top of this hill is reached and the molecule-in-transition falls down the energy hill to form product as new chemical bonds are formed. The structure at the very top of the hill is known as the *transition state* for the reaction. The enzyme must bind the transition state more tightly than it binds either substrate or product. This is an absolute requirement for catalysis. It also provides a terrific way to think about enzyme inhibitor design. If we can come up with a structure that resembles the transition state structure for the reaction, we should have a potent inhibitor. Such molecules are reasonably termed *transition state inhibitors*. This design strategy was used in the successful discovery effort for finasteride, which is described in chapter 6.

Let us imagine that we have identified a target enzyme for which we believe that an inhibitor would find use in clinical medicine. Let us further imagine that, by screening or de novo design, we have found one or more molecules that inhibit the enzyme. Now we need to ask ourselves what needs to be true for such a molecule to have a chance at becoming a drug. It turns out that many things need to be right. One is specificity, which brings us to the matter of secondary screening.

SECONDARY SCREENING SEARCHES FOR SPECIFICITY

Let's go back to our example of the discovery of inhibitors of ACE to treat high blood pressure (p. 53). What we want to do is inhibit ACE, a member of an enzyme class known as *proteases*. However, there are many enzymes in the protease family in the human body; ACE is one. What we need is a *specific* inhibitor, one that targets ACE but leaves all or most of the other enzymes in its class alone. This is the role of secondary screening. Ideally, our molecule will act only on the target enzyme. Because there are many proteases in the human body, this is a large order and it is never possible to be certain that the goal is achieved. Nonetheless, detailed studies are undertaken to be reasonably sure that a high degree of specificity is achieved.[6] Off-target interactions frequently lead to adverse effects in patients. In the specific case of ACE inhibitors, the molecules are very highly specific.

LEAD OPTIMIZATION COMPLETES THE DISCOVERY PROCESS

As mentioned, the process of getting a novel molecule approved for use in human medicine (see Figure 5.2) is generally divided into two stages: discovery and development. Although these overlap some, discovery is generally considered to end when a suitably attractive molecule has been found—one worth pouring a lot of time and money into. This molecule is known as a *development candidate*. Now we need some new vocabulary, and it is included in Table 5.1.

TABLE 5.1

A Vocabulary for Drug Discovery and Development

Term	Definition
Active	A molecule scoring positive for a specified biological activity at a specified potency in a biological assay. At the very outset of a drug discovery effort, the initial objective is to find or design actives, preferably in several structural classes.
Lead	An active molecule meeting a specified set of properties justifying a larger, focused chemistry/biology effort. A lead generally has better potency than the earlier actives, shows significant specificity for the intended target, and may have promising oral bioavailability and a useful duration of action.
Development candidate	A lead molecule meeting all prespecified criteria for advancement to Good Laboratory Practices (GLP) safety assessment. In general, a molecule in this category will be highly potent, highly specific, show good oral bioavailability, have an acceptable duration of action, be resistant to metabolism, have acceptable stability, and has passed preliminary safety studies (generally, in rodents). It may require the synthesis of 1,000 or more molecules to identify a molecule in this category (although there is no guarantee of getting this far regardless of how many molecules are synthesized).
Clinical candidate	A molecule that has passed a rigorous GLP safety assessment, usually in two or three species (typically rats, mice, and dogs). A molecule in this category is ready for phase 1 clinical studies, assuming that appropriate regulatory authorities agree.
Product candidate	A molecule that has progressed to phase 2B clinical trials. A molecule in this category is ready for clinical work to define the dose and dosage regimen to be used during the definitive phase 3 studies.
Product or drug	A molecule that has been approved by appropriate regulatory authorities for marketing.

As mentioned, molecules found to be biologically active in primary screening are known as *actives* or *hits*. As efforts to improve the properties of actives proceed, increasingly attractive molecules are usually found. The following are the properties searched for in increasingly attractive molecules.

- *Potency*: Our actives need to be potent. If they aren't potent, the dose required to effect a useful clinical outcome may be unrealistically large. Many drugs are given as daily doses of a few milligrams to a few hundred milligrams per day for a 70-kg (on average, about 155 lb) human being.

- *Specificity:* I described specificity in my discussion of secondary screening. Bottom line: Find a molecule that does just one thing.

- *Efficacy*: Before testing a product candidate in people, it is necessary to provide evidence that it has the promise of efficacy for the intended purpose. Exploring utility in one or more animal models of disease is the usual course of action. These models have varying levels of reliability as guides to efficacy in people. On the one hand, if an antibiotic for a specific infection works in an animal model, it is likely to work for the same infection in people. At the other extreme, animal models for nervous system disease—anxiety, depression, mania, and so forth—are less reliable than one would desire. Note that potency and specificity in a test molecule do not ensure efficacy for disease treatment. The molecule may not get to the key site of action (for example, the lung in the case of asthma) or it may be eliminated from the body too quickly to act effectively.

- *Safety in animals*: Here, too, one must demonstrate that the product candidate is safe in one or more animal species before initial dosing in people. The extent of required testing depends on the clinical plan. Giving a single low dose to a small number of people requires significant but modest animal safety studies. Giving repeated doses at several levels to people over a long time span requires extensive animal safety studies. Product candidates shown to be unsafe in animal studies simply fail and fall by the wayside. There is no ethical way to determine whether the molecule would be safe in people. Note that safety in animal studies does not guarantee safety in people. If everything else is right, safety in animal studies opens the door to clinical studies, which is where you get the answers to the only questions that really matter: does the molecule work for the intended purpose? Is the molecule safe for the intended purpose?

- *Oral bioavailability*: Oral bioavailability is a measure of the fraction of a dose given by mouth that gets out of the gut and past the liver into the systemic circulation. Most drugs are given by mouth. For many situations, oral administration is the only acceptable route of getting a drug on board. For example, drugs for high blood pressure (hypertension) are taken daily for the lifetime of the patient. Oral administration is the only reasonable route to administer such drugs. Not all molecules taken by mouth find their way out of the gut and into the general circulation. Unless such molecules are intended to treat some disorder of the gastrointestinal tract topically, they are useless. In general, one needs drug molecules that are at least 30 percent bioavailable after oral administration.

- *Distribution*: It is important to know where an administered dose of a drug goes in the human body; this is known as *drug distribution*. For example, a drug for a central nervous system (CNS) disease—say, depression—needs to enter the brain to be effective. In contrast, drugs to treat, for example, a stomach ache should not enter the CNS, where they can do only harm. Drugs to treat asthma, chronic bronchitis, or emphysema need access to the lungs; those to treat infections must gain access to the site of the infection, and so forth. So drug distribution is assessed routinely during discovery and development efforts.

- *Elimination*: Drugs and their metabolites are either eliminated through the kidneys, in the urine, or through the liver, in the feces, or both. It is important to know the route of elimination and it is determined routinely. For example, if a drug is eliminated by the kidneys, special attention must be given to patients who have compromised kidney function. If a drug is eliminated by the liver, special attention must be given to patients with liver disease. These patient populations are studied in clinical investigations and dose adjustments may be recommended.
- *Duration of action*: The optimal dosing schedule for most drugs is once a day. It is easy to remember, convenient, and optimizes patient compliance. No one wants to take a drug every 6 hours, although sometimes this is necessary, albeit inconvenient. Getting a once-a-day or twice-a-day drug requires that it have a long half-life in the body so that efficacy is maintained between doses. The body has several ways of defending itself against foreign molecules, including drugs. It can excrete them via the kidneys (in urine) or via the liver (in feces), or it can metabolize them to (in general) inactive molecules that are then excreted. Getting adequate duration of action can be a big issue for successful drug discovery.
- *Stability*: Many molecules prove unstable over time, and instability can be accentuated by heat, light, or moisture. For drugs dispensed from retail pharmacies, the usual minimum requirement is stability in their packaging for 2 years at room temperature. The stability requirement for drugs dispensed in hospitals may be less demanding, but there will be a requirement.

Clearly, one needs assays for each of these properties. The drug discovery process becomes increasingly difficult and expensive as it proceeds. In the beginning, one needs just one assay and a source of compounds to screen in a search for actives. When found, the actives require the work of medicinal chemists to create better molecules derived from them in terms of the criteria just listed. Now we need a cadre of medicinal chemists to make molecules coupled with a cadre of biologists to run multiple assays. Usually there is also a group of information scientists to help design molecules to be made. A small project quickly becomes large and a major consumer of resources.

As the process proceeds, it is usually the case that a molecule meeting a set of preassigned criteria is created, that molecule is designated a *lead*, and it becomes a focal point of further research. Lead optimization is the stage of work designed to turn the lead into a development candidate (see Table 5.1). The development candidate must meet a longer and more demanding set of criteria for approval, particularly safety. When this is accomplished and a development candidate is approved, discovery becomes development.

Preclinical Development Turns on a Big Machine

Nomination of a molecule to be a developmental candidate is not taken lightly. Approval of such a nomination turns on a big and expensive science machine, figuratively speaking, called *preclinical development*. Large quantities of the development candidate are now needed. A new chemistry effort must be devoted to this cause. Good Laboratory Practices (GLP) assessment of safety in animals will be undertaken, which requires pathologists.[7] Formulation studies are started to determine the optimal format for the molecule in question. The pills you take are not just the effective drug; it is supplemented by any number of other substances that may be required to optimize oral bioavailability, improve palatability, extend shelf life, provide suitability with the demands of high-speed "tableting" machines, and on and on. The degree of oral bioavailability in animals is determined carefully, as is the duration of action. Metabolites are identified and their safety is tested. So resources are consumed and money is spent at an increasing rate, but the best is yet to come.

Clinical Development Is Really Expensive

Assuming that the development candidate passes all safety issues in several animal species and that no other insurmountable barriers arise, the development candidate may be designated a clinical candidate. In the eyes of the pharma house bringing this molecule forward, it has met all criteria for safety and efficacy meriting a clinical trial in people. A huge document is then assembled, collecting all preclinical data, all information on the molecule itself, and proposing a plan for the initial clinical studies. In the United States, this document is called an *Investigational New Drug (IND) application*. If accepted by regulatory authorities—the Food and Drug Administration (FDA), in the United States—clinical studies may begin.

At this point, hundreds or thousands of novel molecules have been synthesized and tested in multiple assays, millions of dollars have been spent, and 2 to 4 years of hard work by numerous scientists have typically passed. Yet, we do not know a single thing about the only things that matter: efficacy and safety in people for the intended use. All the time, work, and money have been spent to justify beginning the process to get answers to these questions. This process is known as *clinical development*.

Clinical development proceeds in phases. At the end of each phase, approval must be sought from regulatory agencies to proceed to the next phase of study. Here is a brief description of each phase.

PHASE I STUDIES INVOLVE HEALTHY VOLUNTEERS AND FOCUS ON
TOLERABILITY AND SAFETY

The first clinical studies are ordinarily carried out in healthy volunteers, not patients. The exception is in the case of drugs for cancer. These drugs are usually highly toxic and ethical issues prevent administering these agents to healthy people. Phase 1 cancer drug studies enroll cancer patients with expected lifetimes of 6 months or less. Phase 1 studies begin at a single site, usually under the supervision of an expert in the relevant field experienced in the very early evaluation of clinical candidates. The primary goal of phase 1 studies is to begin to evaluate clinical candidate tolerability and safety. Tolerability measures include obvious signs of distress, including nausea, headache, pain, and the like. Safety is a more general issue and includes many signs that will not be obvious to a patient, such as abnormalities in blood chemistry, cell or tissue damage, changes in blood pressure or the nature of an electroencephalogram, alterations in kidney or liver function, and on and on. Safety and tolerability are a focal point in all phases of clinical development. The key goal in phase 1 studies, however, is to gain sufficient confidence in drug safety to warrant moving forward to longer term studies involving patients.

One then begins with single, escalating doses of the experimental drug in a small number of volunteers, beginning with one volunteer. As experience is gained and confidence in safety improves, doses are increased and additional volunteers are enrolled in the phase 1 studies. Eventually, the studies move on to multiple doses with corresponding measures of plasma drug levels as a function of time. About 10 percent of molecules entering phase 1 studies eventually make it to the marketplace. Other important attributes to be measured in phase 1 studies include oral bioavailability, duration of action (pharmacokinetics), route of excretion, drug distribution, and metabolism, among others.

PHASE 2 STUDIES FOCUS ON EFFICACY FOR THE INTENDED USE IN PATIENTS

I am going to string together a bunch of adjectives to describe most of phase 2 and phase 3 clinical trials: placebo controlled, randomized, double blind, multiple site, international. *Placebo controlled* means simply that the experimental drug is matched against a molecule known to have neither efficacy nor safety issues (a placebo or "sugar pill"). This is critical because patients may exhibit a placebo effect—that is, they experience an improvement in symptoms in the absence of an effective medication. For example, about 60 percent of patients with depression report improvement in mood and other symptoms of depression while taking a placebo. *Randomization* of subjects entering clinical trials is done to ensure, as far as is reasonably possible, that the group receiving active drug and the group receiving placebo (or a comparative drug) are as nearly comparable as possible; participants are matched for age, weight, gender, race, stage of disease (if any), and so on. In some cases, ethical issues prevent giving clinical trial subjects a placebo; ethically, you cannot put patients at risk by removing them from a drug regimen known to be effective. In these cases, the drug candidate is tested against an approved standard of care, not a placebo. For an example of randomized patient populations, see Table 5.2. *Double blind* simply means that neither the patient nor the physician knows who is getting the candidate drug and who is getting the placebo (or active standard of care). This approach avoids an obvious potential source of

TABLE 5.2

An Example of Baseline Demographic and Clinical Characteristics of Randomized
Patients in a Phase 2 Trial of the Antidiabetic Drug Sitagliptin[a]

Characteristic	Sitagliptin	Placebo
No. of patients	175	178
Age, mean, y	55.6	56.9
Sex, n		
Male	93	103
Female	82	75
Race		
White	127	129
Hispanic	21	22
Black	11	12
Asian	10	5
Body weight, mean, kg	90.9	86.4
Body mass index, mean	32.0	31.0
Duration of diabetes, y	6.1	6.1

[a] The data have been abstracted from a larger table in Rosenstock J., R. Brazg, P. J. Andryuk, K. Lu, P. Stein, Sitagliptin Study 019 Group, 2006. Efficacy and safety of the dipeptidyl peptidase-4 inhibitor sitagliptin added to ongoing pioglitazone therapy in patients with type 2 diabetes: a 24-week, multicenter, randomized, double-blind, placebo-controlled, parallel-group study. *Clinical Therapeutics* 28: 1556–1568. This study was actually conducted on a background of pioglitazone—that is, both sitagliptin and placebo groups were taking pioglitazone (see chapter 12).

bias. Last, multiple investigators usually carry out each clinical trial, and they may be conducted in two or more countries.

Phase 2 clinical studies fall into two categories, unsurprisingly named *phase 2A* and *phase 2B*. Phase 2 studies are carried out in patients, not healthy volunteers. Phase 2A studies assess drug efficacy and continue to monitor safety. Does the clinical candidate work for the intended purpose? Phase 2A studies involve an increased number of patients, use multiple doses, and are carried out at several sites. Opinion leaders in the relevant field of medicine usually supervise these studies. Successful demonstration of efficacy coupled with continued evidence of acceptable tolerability and safety merit progression to phase 2B studies.

A molecule progressing to phase 2B is usually designated a *product candidate*. There is now a reasonably high probability, but by no means a certainty, of achieving approval to market the drug. The goal of phase 2B studies is to define the optimal dose and dosing regimen to be used in the large, definitive phase 3 studies that may follow. Which dose given at which frequency optimizes the efficacy of the product candidate for the intended use while maintaining acceptable patient safety? About 30 percent of molecules entering phase 2B studies eventually become marketed drugs. A successful outcome merits progression to phase 3 studies.

PHASE 3 STUDIES ARE PIVOTAL FOR PRODUCT APPROVAL

Gaining marketing approval for a product candidate in the United States and in many other countries requires appropriate outcomes of two independent, well-controlled clinical studies: phase 3. Phase 3 studies are large, often lengthy, multinational, and expensive. Multiple sites in several countries are usually involved. About 80 percent of molecules entering phase 3 studies become drugs, but not all recover the costs of their development.

PHASE 4 STUDIES BROADEN THE INDICATIONS OR MEANS OF USE OF
AN APPROVED DRUG

Phase 4 includes postmarketing studies designed for a variety of purposes, including gaining approval for new indications, for different doses and dosage regimens, and for alternative dosage forms (a gel capsule, for example) and for alternative routes of administration.

As the clinical program progresses through the various stages, so does the safety assessment program in experimental animals. Justification for administration of single, low doses of a clinical candidate to a human may be gained through safety studies of a 2-week duration in two species, generally in both sexes of rats and mice. As the dose increases, the duration of treatment increases and the number of patients exposed to the candidate molecule increases in the clinical program, as do requirements for safety studies in animals. The duration of treatment increases and additional species may be used. At the end, for a product candidate with the potential to treat patients for long periods, carcinogenicity studies are required, which involve treating mice and rats at the highest dose the animals will tolerate for their entire lifetime, which is 21 months for mice and 24 months for rats. All tissues are examined for evidence of tumor formation.

Each clinical trial involves a multistep process (Figure 5.5). Clinical trial protocols must be approved by regulatory agencies. After clinical investigators have been recruited, the trial protocol must be approved by their home institution. Then, patient recruitment can begin and the actual trial gets under way. Patient recruitment can be problematic and a year or more may elapse between enrollment of the first patient and the last. Trial data must be collected as they are produced, then analyzed, then reported formally. The report is then filed with regulatory agencies as a step in moving to the next clinical trial, which takes time.

Clinical development is nontrivial. The process takes 5 or 6 years and the associated costs may vary from perhaps $100 million to $1 billion or more, depending on the indication for the drug. Clinical development costs for drugs to treat orphan diseases—those afflicting small populations—are generally less.

People who have a low tolerance for failure should find something other than drug discovery and development to do. Scientists and clinicians devote themselves to drug

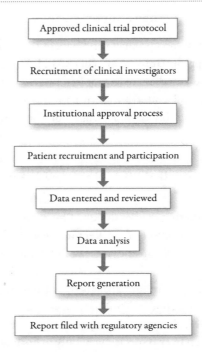

FIGURE 5.5 The Step-by-Step Clinical Trial Process. Successful completion of this process may lead to submission of an NDA (New Drug Application) to regulatory authorities, the Food and Drug Administration (FDA) in the United States.

discovery and development with enthusiasm, given the significance of the work and the potential beneficial outcome for human health around the world. A lot of scientists want to work on really, really tough problems.

The Drug Discovery and Development Process Is Not Linear

I have presented the various stages of drug discovery and development as a series of steps, one following the other. Reality is a bit more complex. Many activities started during drug discovery—before a development candidate is selected—continue through preclinical and clinical development. Perhaps the best example is safety assessment. Have a look at Figure 5.6. Preliminary safety assessment is done before selection of a development candidate. Moving forward, an increasing number of safety assessment studies over increasingly long periods are carried out (some details are provided later in chapter 12). The point is that one does what one needs to do at each stage of discovery and development to ensure patient safety and to optimize opportunities for success at a tolerable cost. This is the case for formulation, scale-up, stability, and metabolism studies as well as those for safety assessment.

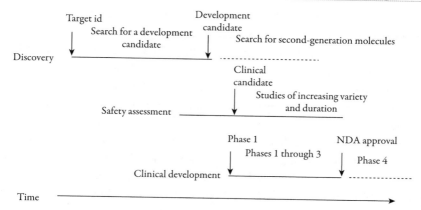

FIGURE 5.6 Development Process. The activities included in discovery, preclinical development and clinical development overlap considerably. The discovery process may be thought of as beginning with target identification or creation of a screen. The culmination of discovery is the identification of a development candidate. But the discovery process does not usually end there as work continues to identify backup molecules should the initial development candidate fail or second generation molecules having improved properties. Safety assessment is an example of a process that begins in late discovery to qualify a molecule as a development candidate and then continues until that molecule either fails or is declared a clinical candidate, and then throughout clinical development with increasing emphasis and resource commitment as the molecule proceeds through phases 1 to 3 and either fails or is approved for marketing. Clinical development does not stop there but generally continues in the search of new indications, novel dosage forms, new dosing regimens, and marketing advantages in phase 4 (post-marketing) studies.

The Final Steps Are Regulatory Approval and Manufacturing

Assuming that the clinical development program has gone well, a pharmaceutical company may collect all data—preclinical and clinical—into a monster document for submission to regulatory authorities. In the United States, this document is known as a *New Drug Application (NDA)*. These documents are submitted to, for example, the FDA electronically. Back in the old days, when they were submitted on paper, the documents were so voluminous the submissions could fill a truck!

The point is to submit an *approvable, marketable* NDA to the FDA. Let's take these issues one at a time.

The quality of the NDA is critical. If the NDA is weak, the reviewing FDA personnel will (rightly) raise a great many issues and questions, Responding to these in a persuasive manner takes time and may require additional clinical studies. These studies will delay, sometimes by years, approval of the NDA. Of course, a truly weak NDA may simply be rejected. It follows that the NDA needs to be of high quality and user friendly—requirements for a rapid approval.

The content of the NDA is also critical. Some of the content is simply required. Beyond the required information, the content of the NDA determines which indications are approved for the drug and which reservations to its use are described in the drug label (a lengthy document intended to tell prescribing physicians how to use the drug). Legally, you can market the drug only for the approved indications. Marketing efforts outside that limit can result in massive fines (hundreds of millions of dollars in some cases) and acute embarrassment to the company. It follows that the NDA must contain data that generate these indications for use that permit the marketing organization to sell the drug. Getting a drug approved that cannot be marketed effectively is a fool's errand.

Approval, never guaranteed, of an NDA can take from a few months—if there is a critical need for the drug and the evidence of efficacy and safety is strong—to a few years, if it is not. While awaiting approval, pharmaceutical houses generally begin manufacture of the product to be able to launch it in the marketplace without delay after approval is granted.

So now you know enough to understand the tales of drug discovery and development that follow. Let's get on with it.

6 Finasteride
THE GARY AND JERRY SHOW

WAY BACK IN chapter 1, I told the story of how I jumped from the world of academic science, which I understood, to that of drug discovery and development, which I did not. So I knew there was a lot of learning to do on my part if I was to contribute anything to Merck Research (and continue to find a way to feed my family), although I underestimated the extent of my ignorance. Here is the way it started.

I showed up for work at Merck Research on January 3, 1979, and met with Ralph Hirschmann, my boss, in his office to get a sense of his expectations and my marching orders. Among other things, he informed me that the benign prostatic hyperplasia (BPH) project was among my responsibilities. At that point, I began to understand just how much I needed to learn. The BPH project had been under way for some years at the time I joined Merck.

Coming in, I knew a lot of general biochemistry and physical organic chemistry. I knew less about human physiology and less than that about human diseases and medicine. Among areas of my ignorance was the disease BPH, for which I was now responsible on the biology side. Later, I became the BPH preclinical team leader.

BPH Is a Significant Medical Issue for Aging Men: Could We Find a Useful Drug?

BPH is medicine's shorthand for benign prostatic hyperplasia, the benign growth of the prostate gland in aging men. Hyperplasia refers to an increased number of cells in an organ, with consequent increase in organ size, as a result of too-frequent cell division.

BPH stands in contrast to the malignant growth of the prostate gland—prostatic carcinoma—a potentially life-threatening cancer. BPH is not life-threatening, but it is surely a condition that compromises the quality of life of men who have it. Lets get started by having a look at what goes wrong.

In normal young men, the prostate gland has a volume of about 20 mL and is roughly the size of a walnut. It surrounds the urethra, the tube leading from the urinary bladder through the penis and through which urine flows from the bladder into the outside world. Have a look at the accompanying illustration of the relevant anatomy in Figure 6.1. Incidentally, women do not have a prostate gland and are therefore free of worries about BPH and prostatic carcinoma. Women have their own gender-specific concerns.

As men age, the prostate gland tends to grow; for example, men in their 70s typically have a prostate gland volume near 40 mL, twice that when they were young. As the prostate gland grows, much of the increase in volume occurs in the worst possible place, around the urethra (Figure 6.1B). This growth tends to constrict the urethra and can lead to a variety of symptoms, such as painful urination, frequency (getting up several times a night to urinate—nocturia—is more than a nuisance), difficulty in emptying the bladder completely, repeated urinary tract infections, and, in extreme cases, acute urinary retention requiring medical intervention.

An enlarged prostate, known as histologic BPH, is very common in aging men. Histologic BPH is diagnosed at autopsy by measurement of prostate size. The percent probability of having histologic BPH is about equal to your age (Figure 6.2). For example, if you are a 70-year-old man, there is about a 70 percent probability you have an enlarged prostate gland. Not all cases of an enlarged prostate, histologic BPH, are associated with symptoms of the disease. Nonetheless, symptomatic BPH is common in older men.

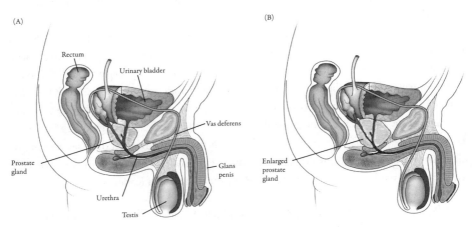

FIGURE 6.1 Male Anatomy and the Prostate Gland. (A) Normal prostate gland. Note that the prostate gland surrounds the urethra, the canal leading from the bladder to the exterior. (B) Enlarged prostate gland. In this view, the consequences of prostate gland enlargement are shown. The enlargement squeezes the urethra, leading to various manifestations of urinary malfunction, including the potential for acute urinary retention, which requires medical intervention.

Whether men seek medical attention depends on the severity of symptoms and the willingness of individual patients to endure them. Minor symptoms are an annoyance, but usually are not a compelling reason to seek out your physician. At the same time, a great many older men do require medical or surgical intervention and, as our population ages, the number will only increase.

Back in 1979, when I became involved in the BPH project, the only treatment for the condition was surgical. There were no drug treatments for BPH available as there are now, a matter to which we return. The usual surgical procedure was known as *transurethral resection of the prostate gland* (TURP). Put in the most basic terms, a device is inserted into the urethra, sited at the prostate gland, and is used to chip away at it to increase the diameter of the urethra. It is not something that I particularly enjoyed thinking about then, and I still do not, although surgical treatment has surely improved in the interim and alternatives are now available.

There were good reasons to search for a pharmaceutical treatment for BPH. In addition to an unpleasant surgical experience, TURP can lead to bleeding problems, and, possibly, urinary incontinence and/or impotence—outcomes that lead to a permanent loss in quality of life. Beyond that, quality surgical care was not, and is not, available universally. To attempt to meet the evident need for a pharmaceutical treatment for BPH, Merck scientists had initiated a search for an appropriate drug. That project had been under way for some years at the time I joined Merck. So I inherited an ongoing, although modest in terms of size, effort.

I want to raise a very general question for drug discovery at this point: given a medical need such as that posed by BPH and one that can be met plausibly by a drug, how do scientists know what to do? You cannot simply get started in the laboratory without some sensible, preferably compelling, plan of research. There are several ways to get to

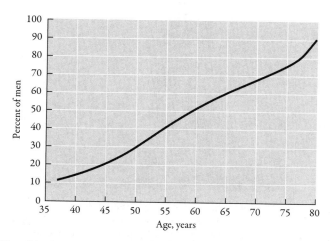

FIGURE 6.2 Plot of the Percentage of Men Showing Evidence of Benign Prostatic Hyperplasia (BPH) as a Function of Age. The data shown here reflect histologic BPH as revealed at autopsy. Not everyone with an enlarged prostate experiences symptoms of BPH, although many will.

sensible plans to pursue drug discovery for a novel medical indication, but usually no truly compelling ones. Sensible plans are frequently not good enough. Many drug discovery efforts are doomed at the outset by one or more factors that were not nor could have been anticipated. Despite decades of intensive research efforts to understand human physiology and biochemistry, we still do not know enough to predict accurately the outcome of pharmaceutical intervention in many instances. The BPH story provides a great insight into one of the most attractive ways of knowing what to do to alleviate

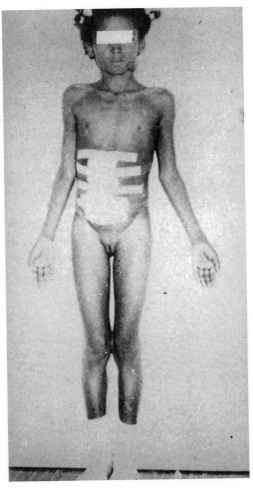

FIGURE 6.3 Photo of a 12-Year-Old Male Pseudohermaphrodite. This child was raised as a girl, one of a small community of such individuals in the Dominican Republic. The external appearance is clearly female and will remain female until the individual reaches puberty, at which time dramatic changes (as illustrated in Figure 6.4) occur. The bandages reflect abdominal surgery for a renal cyst. (Source: Peterson, R. E., J. Imperato-McGinley, T. Gautier, and E. Sturla. 1977. Male pseudohermaphroditism due to 5-alpha reductase deficiency. *Am J Med* 62: 170–91 [photo, p. 175]. Reproduced with permission.)

human suffering: the genetics of human disease. This will take a bit of development, but hang on—the story is compelling.

Male Pseudohermaphrodites Provided an Inspiration for a BPH Drug Discovery Project

The key issue with respect to knowing what to do for the therapy of BPH is to understand the underlying biology: what is the basic cause of the disease at the molecular level? In the story that follows, a single mutation in the gene for a critical enzyme involved in male sex hormone metabolism plays the key role. This knowledge permitted Merck scientists to focus on a specific metabolic event and seek molecules with the promise of setting it right. Here is how the story developed.

Some decades ago, a community of children in a region of the Dominican Republic came to the attention of U.S.-based scientists, notably Ralph Peterson and Julianne

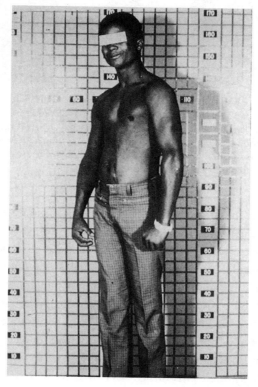

FIGURE 6.4 Same Male Individual with Pseudohermaphroditism at Age 19. This same individual as in Figure 6.3, is now age 19. An individual easily recognized as female becomes recognizably male after puberty. Note the typical male upper body musculature of the individual. (Source: Peterson, R. E., J. Imperato-McGinley, T. Gautier, and E. Sturla. 1977. Male pseudohermaphroditism due to 5-alpha reductase deficiency. *Am J Med* 62: 170–91 [photo, p. 176]. Reproduced with permission.)

FIGURE 6.5 Male Pseudohermaphrodites. The male on the left and the male in the center are pseudohermaphrodites and are cousins. The male on the right is a normal male and the brother of the affected male on the left. The affected males are more muscular, whereas the normal male has a beard and temporal hairline recession. (Source: Peterson, R. E., J. Imperato-McGinley, T. Gautier, and E. Sturla. 1977. Male pseudohermaphroditism due to 5-alpha reductase deficiency. *Am J Med* 62: 170–91 [photo, p. 178]. Reproduced with permission.)

Imperato-McGinley. The latter is now chief of the Division of Endocrinology, Diabetes, and Metabolism at the Weill Cornell College of Medicine in New York City.[1] At birth, these children are sexually ambiguous but are recognized as girls and are raised as girls. To get an idea of pseudohermaphroditism, have a look at the 12-year-old individual in Figure 6.3, recognized and raised as a girl. At the time of puberty, this individual underwent a truly remarkable transformation. See Figure 6.4 for an after-puberty comparison of the same individual. Specifically, a penis develops, a scrotum develops and testicles descend into it, the typical upper body musculature of males develops, the voice deepens, a typically male pattern of hair develops, and so on.

Perhaps surprisingly, these girls-turned-boys are heterosexual—that is, they pursue girls as romantic targets despite having been raised as girls. Beyond that, they exhibit a certain macho characteristic (Figure 6.5). Technically, these individuals are termed *male pseudohermaphrodites*. One such individual is the protagonist of Jeffrey Eugenides highly successful book *Middlesex: A Novel*.[2]

Clearly, something is seriously amiss here. Although cases of ambiguous gender at birth do occur rarely, there is nothing obviously ambiguous to the casual observer about these people at birth. The extraordinary change at the time of puberty is totally unexpected, dramatic, and very rare indeed. These male pseudohermaphrodites are victims of a single-gene genetic disease. Specifically, they have a defect in an enzyme known as *5α-reductase.*

Although it was not known at the time, there are two 5α-reductases coded for in the human genome; the one of primary interest for this story is known as *5AR2.* It is localized to skin, seminal vesicles, the prostate, and the epididymis. The other, 5AR1, is localized in the hair follicles and brain.

As we shall see, this defect in 5AR2 accounts for the apparent sex change at puberty of affected individuals. It also provided a strong rationale for our drug discovery effort. We need to take this one stage at a time. Let's begin with the role of male sex hormones, androgens, in development *in utero.*

Male Sex Hormones Work Through the Androgen Receptor

The role of 5AR2 in metabolism is to convert the male sex hormone, or androgen, testosterone (T) into a closely related molecule, dihydrotestosterone (DHT; Figure 6.6).

Of the male sex hormones, DHT is the more potent, about 10 times stronger than T. These sex hormones do their magic by interacting with an androgen receptor and, in the final analysis, controlling several aspects of protein synthesis.

There are three keys to understanding male pseudohermaphroditism. First, we all start out life *in utero* as females. In the case of a fetus that is destined to be a male, there is a surge of androgen activity, T and especially DHT, between weeks 9 and 17 of gestation. This surge causes female structures to degenerate and male structures to form. Second, as noted, DHT is a more powerful androgen than T. Third, affected individuals have a defective 5AR2. Let's look closely at the logic here.

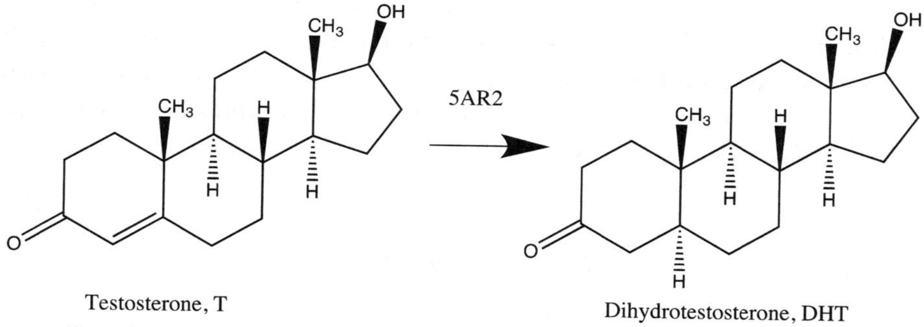

Testosterone, T   Dihydrotestosterone, DHT

FIGURE 6.6 Conversion of the Male Sex Hormone Testosterone (T) into a Closely Related Molecule, Dihydrotestosterone (DHT). This reaction is catalyzed by the enzyme 5AR2, the molecular target for discovery of drugs for benign prostatic hyperplasia.

A defective 5AR2 means that the conversion of T to the more powerful male sex hormone DHT is compromised in affected individuals. Indeed, the ratio of T to DHT in the blood of these males is substantially larger than in men with an intact 5AR2. DHT levels are too low and T levels are too high, which is the expected result of a compromise in the rate of conversion of T to DHT. The lack of normal levels of the potent androgen DHT between weeks 9 and 17 of gestation results in incomplete degeneration of the female structures and incomplete formation of the male structures at the time of birth. So the newly born males look typically female and are recognized by their parents as girls.

After birth, there is little difference in male hormone levels between boys and girls until the time of puberty. At that time, males get a surge of androgens, T and DHT, which is largely missing in females (females do have low levels of androgens). This androgen surge results in deepening of the male voice and development of the upper body musculature of men. In the specific case of males lacking a fully functional 5AR2, this androgen surge completes the differentiation process that normally occurs *in utero*.[3] The androgen surge at puberty, mostly T in these men, makes up for the lack of adequate androgen exposure *in utero* resulting from the deficiency of DHT. Children who are apparently females turn into apparent males (refer back to Figure 6.4).

So what is wrong with 5AR2 in these men? Although the answer is not known in detail, there has been an error in the amino acid sequence of the 5AR2 protein or a deletion of the segment of amino acids in the protein. Remember that the amino acid sequence determines the three-dimensional structure of the protein and, in turn, the biological activity of the protein depends on the three-dimensional structure. So an error in sequence may translate into an error in function. In our case, the result is a less than fully functional 5AR2. The consequence is male pseudohermaphroditism. This is a good example of a human genetic disease caused by a defect in a single protein-coding gene. There are many such genetic diseases known, including sickle cell anemia, cystic fibrosis, Huntington's disease, Fragile X syndrome, and on and on.

So why is this genetic disease of interest for drug discovery? Here is the point. *These individuals have small prostates and they never grow, they do not get male-pattern baldness, and they do not get acne.* The simplest explanation for these observations is that DHT is the causative agent for the benign growth of the prostate, BPH, for male-pattern baldness, and for acne. It follows that if we could somehow recreate the diminished levels of DHT in men who are otherwise healthy we might be able to prevent (or cure) BPH, acne, and hair loss.

Turning Inspiration into Drug Discovery Is Not So Easy

The next question is: how do we go about recreating a deficiency of DHT in healthy men? A related question is: what are the consequences of doing this in terms of potential adverse effects? We start with the first issue: finding a way to decrease DHT levels in normal men.

The point is to find some way of reducing the activity of 5AR2. Nature does it by making a nonfunctional or less than fully functional enzyme. Scientists can do it by designing and making an inhibitor of the enzyme. As mentioned earlier, enzyme inhibitors are generally small molecules that occupy a site on the enzyme that renders it unable to interact with the normal substrate and convert it to product. In our case, the normal substrate is T and the product is DHT. Our focus is how to find a molecule that will inhibit 5AR2 effectively and specifically.

There are several ways for searching out enzyme inhibitors: screening of compound libraries, screening of natural products, designs based on detailed three-dimensional structures of the target enzyme, and designs based on the structures of substrates, products, or something intermediate between them, known as a *transition state*. For the search of inhibitors of 5AR2, the last option was selected. The BPH drug discovery project was small in terms of manpower. The screening of compound libraries and natural products tends to be manpower intensive, so these options were unattractive. The detailed three-dimensional structure of 5AR2 was unknown at the time (and still is), ruling out structure-based drug discovery. In contrast, design of inhibitors based on the structures involved in the reaction seemed promising for this project. Merck chemists set about designing potential inhibitors of 5AR2 based on the structures sharing features of T and DHT: transition state inhibitors.

This strategy turned out to be the way to go. The chemists involved designed a clever mimic of the transition state for the conversion of T to DHT. The fundamental idea was to start with T and introduce a nitrogen atom into the first ring of the steroid structure and to put a new hydrogen atom where it is in DHT (Figure 6.7).

They then synthesized several molecules around this design and tested them as inhibitors of 5AR2. The design proved to be inspired, the novel molecules worked as inhibitors, and we now had a good starting point for the discovery work. The early inhibitors lacked many of the properties required to be designated a development candidate:

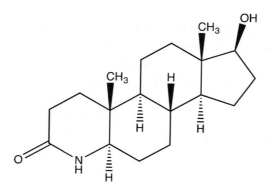

FIGURE 6.7 The Structure of a Concept Molecule for 5AR2 Inhibitor Design. This structure has the properties of a structure, the transition state, on the pathway from testosterone (T) to dihydrotestosterone (DHT). It is a concept transition state inhibitor of 5AR2.

potency, specificity, oral bioavailability, long duration of action, efficacy in an animal model of BPH, and safety. However, the point is that we had a starting point, something for chemists, in collaboration with biologists, to work from in an effort to build in all the necessary properties of a drug. So that exploration began.

Chemists and Biologists Work Together in Drug Discovery Projects

It is time to introduce the people who were responsible for making this happen. Gary Rasmusson led the chemistry effort. Gary had a modest number of people working for him making molecules. Gary reported to Burt Christensen, Vice-President for Chemistry, at the Rahway, New Jersey, site of Merck Research. Jerry Brooks led the biology effort. Jerry had just two scientists working for him—Charlie Berman and Ray Primka—and doing all the necessary assays. Jerry initially reported to me and later to Eve Slater, who was recruited to lead the endocrinology department (and who later became a Merck Research senior vice-president).

Here is the way that it went. Jerry and his people had rigged an assay that determined the potency of the molecules that Gary and his people created in the chemistry lab as inhibitors of 5AR2. The data that Jerry generated were returned to Gary who used them to design the next series of molecules to be made, with the point being to follow a path to greater potency. Doing this is a little like finding your way through a maze by trying one route after another. You take one direction, find that it is a dead end, reverse your path, choose another route, find that it looks promising, and continue. As you find your way through the maze, you make a large number of wrong turns, but little by little, learning from your mistakes, you find increasingly promising ways through it. Finding a molecule good enough to be a clinical candidate corresponds to getting out of the maze. Failing to do this corresponds to wandering in the maze forever (in the pharmaceutical industry, you are eventually freed from the maze by management killing your project and assigning you to a new maze).

As Gary and his people created molecules with promising potency, a new issue arose. Are these molecules just good enzyme inhibitors in the test tube or will they actually inhibit the enzyme in a living animal? To answer this question, Jerry needed a new assay—how to measure enzyme inhibition in an animal. For reasons developed next, the animal of choice was old beagle dogs. The measurement was the ratio of T to DHT, T/DHT, in the plasma of these animals before and after treatment with varying doses of 5AR2 inhibitors. Because our goal was to make a drug that could be taken orally, these inhibitors were fed to the beagles in a small capsule hidden inside a meatball. Effective inhibition of the enzyme *in vivo* was revealed by increases in the T/DHT ratio in plasma samples taken from the dogs (remember that the goal is to maintain the concentration of DHT low relative to that of T).

Doing the test tube assays is straightforward. Measuring the extent of enzyme inhibition in beagles is far more difficult. In the first place, you need a great deal more material to feed to beagles than you need for the test tube assay. Second, serial blood samples from the beagles must be taken and, last, the blood samples must be assayed for T and DHT, which is not so easy, and Jerry and Gary had to agree on a set of criteria for advancing test inhibitors to this stage. Suffice it to say that the chemistry work did produce a family of molecules found to be effective in inhibiting 5AR2 in beagles.

But we were not done. The point was not to make great inhibitors that work in living beagles but to make molecules that would be useful in the treatment of BPH in men. So, we needed an animal model for the human disease. It happens that aging male beagle dogs do get an enlarged prostate, just as aging men do. So old beagles with enlarged prostates served as our model for human BPH. (I have heard it said that old male lions develop enlarged prostates, and this may be so, but lions make challenging experimental animals.)

Most animal models of human diseases are not perfect. As I emphasized earlier, the similarity in biochemistry and physiology among living organisms is spectacular. If it were not, we would not have animal models of disease. At the same time, there are obvious differences in physiology that can compromise the reliability of animal models of disease in predicting the outcome in people. Our animal model for BPH was useful but not perfect. Here is the principal point. The prostate gland of dogs has a great deal of soft glandular tissue and rather little tough stromal tissues. The human prostate is just the opposite—rich in tough stromal tissue and poor in glandular tissue. So we knew in advance that we could not hope to see the degree of change in the prostate of human BPH patients that we might see in beagles. It is far easier to shrink glandular tissue than stromal tissue. However, the beagle model of BPH was all that we had, so we used it, cognizant of its limitations.

The most effective inhibitors that Gary and his group made judged by the *in vivo* test of enzyme inhibition were selected for examination for their ability to shrink the prostates of beagles with enlarged ones. Here, again, the resources required for the assays had increased. Dogs were dosed once daily with inhibitors given orally, and prostate size was measured before treatment and again after treatment for 6 weeks.[4] We had to choose carefully which molecules were tested. These are long-term experiments in living animals, and substantial amounts of material and dedicated manpower were required.

The results were spectacular. The 5AR2 inhibitors were remarkably effective in reversing BPH in beagles. In favorable cases, prostate glands would shrink by a factor of four—that is, a prostate that occupied a volume of 20 mL before treatment would shrink to 5 mL after treatment. A gland the size of a walnut was reduced to one the size of a peanut.

So here is what we knew. Gary's inhibitors were potent inhibitors of 5AR2 in the test tube, potent inhibitors of the enzyme in beagle dogs, effective given orally, had a long duration of action because they were effective given once a day, and were powerful agents for shrinking enlarged prostate glands in beagles. Jerry's assays coupled with Gary's medicinal chemistry had generated a family of highly attractive agents for BPH. The remaining issue before finding a molecule suitable for clinical studies in people was safety in animals.

Safety Is the Biggest Hurdle to Overcome in Getting to a Drug

At the doses found to be effective for shrinking the prostates of beagles, we saw no adverse effects. The dogs seemed well and happy. However, before moving a molecule into the process required for clinical trials, more stringent measures of safety are required. We chose the best of our inhibitors for rigorous safety assessment. It did not have a name but it did have a number—MK216 (Figure 6.8):

Modest but critical structural changes had been made to the molecule selected as a starting point, which may seem like an easy thing to do, but this is not the case. It is worth comparing the two structures. There are an almost indefinite number of ways to alter the structure of a starting molecule and there was no way to know which of them would be likely to yield a molecule with improved properties. It is remarkable that Gary, with a handful of chemists, and Jerry, with his two biologists, were able to create and evaluate molecules with a highly promising set of druglike properties for treatment of BPH.

MK216 was examined for safety in a number of careful studies. These studies were carried out by a unit within Merck Research specifically tasked with determination of safety issues; we called it *Safety Assessment*. During these preliminary safety assessment studies, a number of species were exposed to high levels of MK216 given orally for various time periods, and signs of toxicity were sought. Note that regulatory authorities require that evidence of toxicity in animal studies be demonstrated, which vary from simple observations such as vomiting, signs of pain, abnormal behavior, and the like, to measurements

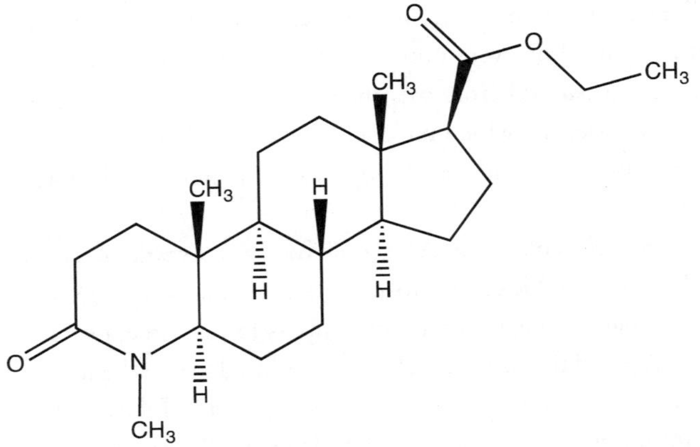

FIGURE 6.8 The Structure for MK-216, a Potent and Specific Inhibitor of 5AR2. MK-216 proved highly effective in shrinking the enlarged prostate glands in beagle dogs. It is worthwhile to compare this structure with that provided in Figure 6.7, our starting point. The changes made are small but highly important in terms of effective enzyme inhibition.

of heart rate, blood pressure, blood enzyme levels, and other quantitative readouts of health. Last, treated animals were euthanized and a dozen tissues were examined microscopically for evidence of damage. Here are the results for MK216. The molecule was found to be safe in rats and mice, and safe in rabbits and chimps in short-term studies, but hepatotoxic (toxic to the liver) in dogs at doses considerably higher than those required to shrink the prostate of old beagle dogs.

So we have the following situation. A highly promising molecule, the result of several years of work by Merck scientists, is found to be safe in several species but not in dogs at high doses. The question is whether this molecule is suitable for approval for development, possibly leading to clinical trials in people. Development includes all those activities required for approval for clinical studies as well as those studies themselves (see chapter 5). Approval for development turns on the big science machine. Development chemists must make large quantities of material, pharmaceutical chemists must come up with a suitable formulation (drug plus other substances required for good oral bioavailability, stability, and ease of administration) for formal safety assessment studies, the safety studies must be carried out according to GLP standards (see chapter 5), clinicians begin to create a strategy for clinical studies, statisticians consult on how many patients may be required to get a reliable estimate of efficacy, drug metabolism gears up to measure plasma drug levels as a function of dose and time, and on and on. So an affirmative decision is not taken lightly. On the other hand, to reject a molecule that shows high promise of becoming a marketed drug and benefiting patients worldwide would be a disaster. What did Merck management decide to do with MK216?

Merck elected to reject MK216. Patients would have to take a drug for BPH long term. There were legitimate worries that, even at a low dose of MK216, sooner or later liver toxicity would show up in people. So Gary and Jerry were not out of the maze. Their next

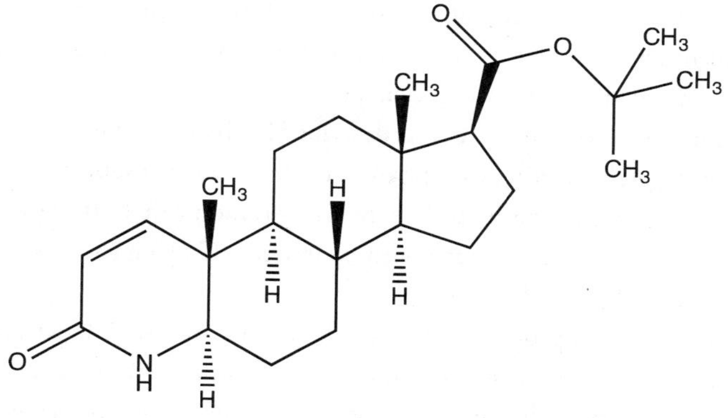

FIGURE 6.9 The Structure for Finasteride, a Potent and Specific Inhibitor of 5AR2. Finasteride proved to be safe in multiple safety assessment studies. It was approved for development by Merck management. Compare this structure with that of MK-216 in Figure 6.8, a failed molecule for reasons of toxicity in dogs.

task was to find a molecule that had all the attractive properties of MK216, but was safe in dogs and other species as well.

This turned out to be a huge job because it involved one additional difficult assay: hepatotoxicity in dogs. Doing these studies requires large amounts of material and lots of time for the safety assessment. To condense a long story, it required an additional 3 years of effort to come up with a small family of molecules that met our criteria. We thought that one of them—it became known as *finasteride*—was the best of the lot (Figure 6.9).[5]

Finasteride differs in only small ways from MK216, but these modest differences make all the difference: increased potency and safety in experimental animals. This molecule was approved for development, which cheered up those of us who had worked on the project for years.

Finasteride Becomes a Marketed Product for BPH

Finasteride passed the initial safety assessment hurdle. We now had all the data we required to file an Investigational New Drug application (IND), with the FDA. The IND basically contains all information on finasteride that qualifies it for initial clinical trials; the focus is on safety in animal studies and efficacy in an animal model of BPH. It includes the phase 1 clinical trial protocol. Once submitted, the FDA had 30 days to reject the IND. If they did not, Merck was free to begin clinical trials, which was the case for finasteride. We were free to begin. Clinical trials are designed to answer the only two questions that matter for approval for marketing: is the drug safe for the intended use? Is it effective for the intended use?

The BPH Discovery Project Team was reconstituted to form a Development Project Team with a clinician at its head—Elizabeth Stoner, a highly capable physician/scientist who was eventually promoted to senior vice-president in Merck Research Laboratories. Elizabeth replaced me as team leader and Jonathan Tobert, a clinician who worked with the BPH Project Discovery Team. Jonathan went on to lead the Development Project Team for mevastatin and lovastatin (see chapter 8).

Phase 1 clinical trials were initiated. As described in chapter 5, clinical studies proceed in a number of stages, beginning with phase 1 studies in normal subjects and proceeding to more advanced studies in patients. Patient safety is monitored throughout; efficacy is determined after phase 1 studies have been completed. Getting the dose right and the frequency of dosing right are key issues.

As the clinical program progresses through the various stages, so does the safety assessment program. Justification for single, low doses of a product candidate to a human may be gained through safety studies of a 2- to 4-week duration in two species, usually in both sexes. At the same time, requirements for safety studies in animals increase—larger number of animals, perhaps new species, longer duration of treatment—as required to justify higher doses, multiple doses, dosing for longer periods, and an increasing number of patients exposed to the product candidate in the clinical program. Last, for a product candidate with the potential to treat patients for long periods, carcinogenicity studies are required, which includes exposing mice and rats to the highest tolerated dose for their lifetime. All tissues are examined for evidence of tumor formation.

In the case of finasteride, here is how events unfolded.[6] Phase 1 trials in healthy volunteers went smoothly with no evidence of toxicity. Efficacy was demonstrated in phase 2A trials. Increases in the T/DHT ratio in human plasma established that finasteride was effective in inhibiting 5AR2 *in vivo*. Phase 2B trials established that a 5-mg dose taken orally once daily was appropriate for phase 3 studies. Phase 3 studies were extensive. They included two 1-year randomized, double-blind, placebo-controlled studies. These studies were extended in the form of a 5-year open-label study (that is, the blinding of patients and physicians regarding who was getting the drug and who was getting the placebo was lifted). In addition, a randomized, double-blind, placebo-controlled 4-year study enrolling 3,040 patients was completed. These studies established that finasteride, taken over time, reduced the incidence of acute urinary retention, reduced the need for BPH-related surgery, increased urinary flow rate, and reduced the enlarged prostate gland in patients with BPH by about 20 percent, with relief of the symptoms of BPH. Throughout, the safety of finasteride was monitored. It was found to be safe for use in the treatment of BPH. The FDA approved finasteride for marketing in 1992, which was the hallelujah moment for finasteride. Proscar was chosen as the trade name. A generic version of finasteride is now available.

Finasteride works as we had hoped, but not as rapidly as we had foreseen. Relief of symptoms generally requires a few months of treatment, as much as 6 months for some patients. As always for drugs, there are some side effects associated with finasteride use.

As the clinical trials progressed and the probability of success improved, Merck made plans for the large-scale manufacture of finasteride itself and the final dosage form that would be marketed. The final dosage form includes a small group of excipients as required to provide product stability, tablet strength, and consistency with high-speed tableting machines.

Clinical efficacy and safety studies, however, do not end with approval of the first indication for a drug. Studies continue to explore more fully the biological profile of the drug, to search for new uses and indications, to evaluate the drug in special populations, to respond to issues that may arise with widespread use, to explore the utility of new dosage forms or routes of administration, and to assist in marketing the drug.

Finasteride Reduces the Prevalence of Prostate Cancer in Men

The benign growth of the prostate gland in aging men causes BPH. The abnormal, malignant growth of the prostate gland is prostate cancer, a common tumor in older men. Given the favorable effects of finasteride on the prostate gland found in the BPH clinical program, perhaps finasteride would prevent, reduce the prevalence, or delay the onset of prostate cancer in men.

After the approval of finasteride for BPH, the Prostate Cancer Prevention Trial addressed this issue. A total of 18,882 men with *normal* prostates were treated with finasteride 5 mg once a day for 7 years in this randomized, double-blind, placebo-controlled study. The study was completed and published in the *New England Journal of Medicine* in 2003. The study revealed a 25 percent reduction in the prevalence of prostate cancer in this group of men—a heartening result. Less heartening was the observation of a small but significant increased risk of development of a high-malignancy form of prostate cancer. It is not clear whether this aggressive form of prostate cancer is a consequence of finasteride treatment. Perhaps finasteride treatment unmasks the presence of an aggressive form of cancer already present. Although long-term use of finasteride reduces the likelihood of developing prostate cancer, it is not used regularly for this purpose. The slightly increased likelihood of developing a particularly dangerous form of the disease, regardless of whether it is the result of finasteride use, has discouraged such use. The Prostate Cancer Prevention Trial cost Merck hundreds of millions of dollars. Despite the success of the trial and the approval of finasteride for the indication of prevention of prostate cancer, the cost of the trial was not recovered. Pharmaceutical research is a risky, high-stakes business and not for the faint of heart.

Inhibition of 5AR2 Is Not the Only Pharmaceutical Treatment for BPH

Merck was not the only pharmaceutical house focused on finding a drug to treat men with BPH. At the time, Merck was the only pharmaceutical house exploiting inhibition of 5AR2 for this purpose. A decade later, GlaxoSmithKline developed dutasteride, marketed as Avodart, which is related structurally to finasteride but is capable of inhibiting both 5AR1 and 5AR2.

Others elected to bring forward a class of molecules known as *alpha blockers* to treat BPH. Alpha blockers act at their receptors to relax smooth muscles in the bladder, facilitating urination. Two examples include doxazosin, developed by Pfizer and marketed as Cardura, and tamusolosin, developed by Yamanouchi Pharma in Japan and marketed as Flomax. Although they have no effect on the prostate gland itself, they are effective in improving the symptoms of BPH.

Because 5AR2 inhibitors and alpha blockers act by distinct mechanisms, you might imagine that combining the two would improve the outcome for BPH patients. This was the case. Merck carried out a 4- to 6-year study in which the combination of finasteride and doxazosin was compared with finasteride alone and doxazosin alone in 3,047 men with BPH. The combination was found to be significantly superior to either agent used alone. GlaxoSmithKline now markets a combination product of dutasteride and tamusolosin known as *Jalyn*.

New Applications: Finasteride Has Potential Use for Male-Pattern Baldness and Acne

We need to get back to the case of the male pseudohermaphrodites that began this story. As I noted earlier, these men have a small prostate that does not grow (the inspiration for our BPH program), they do not get male-pattern baldness, and they do not get acne. Other than their urogenital issues, they are healthy. So would an inhibitor of $5AR_2$ also prevent male-pattern baldness and acne? Let's consider these possibilities one at a time, beginning with male-pattern baldness.

By some chance, I met Roger S. Rittmaster, a physician and endocrinologist at the National Institutes of Health (NIH) at the time, and this provided a way to look at finasteride for male-pattern baldness. The key was that Roger had access to a family of stump-tailed macaques. As far as I am aware, these are the only primates other than man that go bald spontaneously. So, they might serve as a useful animal model of male-pattern baldness. Roger and I elected to work together to determine whether we could generate some encouraging data.

Our approach was very simple. We made a cream containing finasteride suitable for smearing on the heads of the monkeys. They were treated daily for some time. Our assay for efficacy was simplicity itself: Roger shaved the macaques heads from time to time and weighed the shavings. The results were clear. In those monkeys treated topically with finasteride, the shavings were heavier than in the control, untreated group. Heavier shavings meant less hair loss—more to shave off. The effect was also clear visually. Treated monkeys were visibly less bald than the untreated monkeys.

So we had an apparently clear path for the development of finasteride for male-pattern baldness. The monkey data strongly suggested that it would work, and we had safety data in men from the BPH studies. What could possibly go wrong?

There is a ticklish ethical issue here. Given the urogenital defects in the men with pseudohermaphroditism, it seemed certain that finasteride, and all other $5AR_2$ inhibitors, would be teratogenic for male fetuses—that is, they would create birth defects in males, likely similar to those in men afflicted with defects in the $5AR_2$ enzyme. Now it is surely true that a drug for men only, handled properly, has no potential for creating birth defects. Men do not become pregnant.

However, there is an issue that I have not yet brought up: finasteride will also work for treatment of female hirsutism—the growth of hair in the male pattern in women. Contrary to what you might expect, based on the discussion thus far, $5AR_2$ inhibitors will prevent male-pattern hair growth in women. Hirsutism is a significant cosmetic issue for many women; take a look at the number of hair removal clinics for women in your phonebook or on the Internet. It follows that there is significant risk that a product on the market for male-pattern baldness would also be used to treat female hirsutism, which is an example of off-label use. Most women afflicted with hirsutism are postmenopausal and cannot become pregnant. But, there is a community of women capable of becoming pregnant who have female hirsutism.

So we had to balance two factors. On the one hand, we had a plausible treatment for male-pattern baldness, a cosmetic issue that compromises the quality of life for many men. On the other, there was the risk of off-label use by women of child-bearing potential with a risk of creating birth defects. When I proposed developing finasteride for male-pattern baldness, Merck management rejected the idea based on worries about off-label use for female hirsutism.

At some later date, after I had left Merck for another opportunity, Merck management reversed their position and chose to develop finasteride for male-pattern baldness. Merck undertook a 5-year study of the efficacy of finasteride, given 1 mg/day orally, for the prevention of hair loss in men. On average, two of three treated men regrew some hair whereas subjects in the placebo group continued to lose hair. Finasteride works for this indication, and the FDA granted marketing approval in 1997, another hallelujah moment. Propecia is the trade name for finasteride in 1-mg tablets for prevention of hair loss in men.

The remaining issue is acne. Should finasteride be developed to prevent or treat acne? The ethical worries for use of finasteride for prevention of hair loss are clouded, and reasonable people can arrive at opposite conclusions. However, the situation for acne seems quite clear. Acne is largely a cosmetic issue for the young. Female hirsutism is largely a cosmetic issue for women beyond their child-bearing years. Young girls get acne and they have child-bearing potential. It is impossible to believe that an acne treatment for young men would not also be used by young women. The risk of birth defects resulting from treatment for a transient cosmetic issue is clear and compelling.

Beyond that, there was and is an available treatment for severe, scarring acne: Accutane. This drug is highly teratogenic and it must be used only in the context of proper birth control. No one that I know thinks that we need another teratogenic drug to treat acne. This is the end of this part of the story.

In summary, an uncommon genetic disease, an error in the gene that codes for 5AR2, provided inspiration for the discovery of finasteride and, later, dutasteride. Much of the credit for the Merck drug surely belongs to Gary Rasmusson and Jerry Brooks, who saw this effort through to completion over many years, and to Elizabeth Stoner, who directed a splendid clinical effort. It seems to me, a great story.

Gary Rasmusson, Jerry Brooks, and Elizabeth Stoner were awarded Merck's highest honor—the Director's Scientific Award—for their efforts on finasteride.

7 Basic Research, Snake Venoms, and ACE Inhibitors
ONDETTI, CUSHMAN, AND PATCHETT

ON APRIL 6, 1981, the U.S. FDA approved a molecule *captopril* for the treatment of elevated blood pressure, otherwise known as *hypertension*. This action marked the beginning of a major change in the treatment of high blood pressure and, later, congestive heart failure, diabetic kidney damage, postheart attack management, and stroke prevention. The health of millions of people worldwide has been sustained or restored through the use of captopril and its pharmaceutical relatives.

Captopril is the product of research and development work at the Squibb Institute for Medical Research, at the time a branch of Squibb Pharmaceuticals, now Bristol-Myers Squibb. Captopril was marketed worldwide under the trade name Capoten. Its patent life has long since expired and it is available as a generic drug.

Captopril Was a Breakthrough, Pioneer Drug

Captopril is remarkable for two reasons. First, it is the pioneer drug in a new class of treatments for high blood pressure and other cardiovascular problems. This new class is known as ACE inhibitors. ACE is the chemical acronym for angiotensin converting enzyme. The names of enzymes reflect an effort to describe what they do, and the names are often complicated. Later, I explain what ACE does, and perhaps the name will make sense then. In the meantime, ACE will have to do.

Second, captopril was among the first drugs approved for use in clinical medicine discovered by a process of rational drug design, as opposed to a more-or-less random hunt for molecules having promise of use in human medicine. The discovery of captopril followed the earlier work of James Black, Gertrude Elion, and George Hitchings, pioneers in the formulation of the principles of drug design and treatment. They shared the Nobel Prize for Medicine or Physiology in 1988.

The goal of this chapter is to pull together the various threads of the story of ACE inhibitors and weave a revealing fabric.

Treatment of High Blood Pressure Has a Long History

Physicians have been measuring blood pressure for a lot longer than they have had effective means to deal with excursions beyond the normal range. It has long been known that high blood pressure is a significant risk factor for heart attack and stroke. Former President Franklin Delano Roosevelt had a decade-long history of high blood pressure when he was felled by a stroke on April 12, 1945. So did Joseph Stalin, who died of a stroke in 1953. The pharmaceutical industry recognized that there was a critical medical need to be filled and value to be created through finding safe and effective drugs to control blood pressure.

Patients with high blood pressure do not feel ill. Absent other problems, they feel well. Furthermore, there are no drugs that cure high blood pressure; they control it but they do not cure it. It follows that medications for high blood pressure, once recognized, need to be taken for a lifetime. This fact puts some pretty strict limits on the set of acceptable properties for blood pressure medications. They need to be effective, well-tolerated and safe, and convenient to take.

The earliest efforts to control blood pressure advocated restriction of salt intake, which remains good advice today. However, for most people with high blood pressure restriction of salt intake, although helpful, is inadequate to achieve acceptable control. More help is required.

The first set of drugs used to reduce blood pressure fell well short of an acceptable profile. They were based on a toxic element and their adverse effects were inconsistent with a decent quality of life for patients with high blood pressure. A breakthrough occurred when Merck scientist Karl Beyer discovered the diuretic known as chlorothiazide (marketed as Diuril). Diuretics promote the excretion of water from the human body, reducing blood volume and, therefore, pressure. Several other safe and effective diuretics were discovered subsequently. Diuretics remain a mainstay for treatment of high blood pressure and heart failure. The most widely used diuretic is hydrochlorothiazide, a simple derivative of the molecule discovered by Beyer. It is marketed as HydroDiuril, among other names.

The discovery of safe and effective treatments for high blood pressure is a pivotal point in the practice of medicine in support of human health. For years, people only sought out their physician when they felt ill. After the health benefits of blood pressure lowering were established through clinical studies, people feeling well sought out their physicians for blood pressure measurements (now readily done at home). Beyond that, physicians began to check blood pressure routinely for their patients, regardless of the reason for their visit. Prevention of the consequences of high blood pressure—heart attacks, strokes, kidney failure—was the goal. It is easier to sustain good health than to restore it.

Pharmaceutical companies do not enjoy being shut out of an exciting new class of treatments for human diseases, so it should come as no surprise that a number of major pharma companies jumped on the Squibb discovery of captopril with the intention of finding safer or more effective molecules. Merck was foremost among them. In fact, the launch of captopril was followed by two Merck ACE inhibitors: enalapril (marketed as Vasotec and Renitec) and lisinopril (marketed as Zestril and Prinivil). Note that all the generic names end in *pril*. It is common in the pharma industry to have all drugs sharing the same mechanism have the same suffix. ACE inhibitors are "prils." Other prils followed the Squibb and Merck drugs. As the number of ACE inhibitors increased, so did the scope of their use. As noted, they are now used for treatment of heart failure, kidney protection, to improve the outcome for heart attack victims, and to lower the incidence of strokes. As a class, ACE inhibitors have had a profound, beneficial effect on human health.

Let's give thought to how these drugs work their magic.

The Renin–Angiotensin–Aldosterone System Is a Key to Blood Pressure Control

The physiological control of human blood pressure is complex. In its intimate details, it is not fully understood. However, much is known, and that knowledge has been put to excellent use. It serves as the foundation for blood pressure control.

One of the central systems for blood pressure control—and a focal point for multiple blood pressure control drugs including captopril—is the renin–angiotensin–aldosterone (R-A-A) system. Recall that we had a pretty close look at the structure of the enzyme renin back in chapter 3. Here we are going to explore what it does. However, before we get up close and personal with the three actors in the R-A-A system, let's pull together a simple image of what affects human blood pressure.

Think of the human vascular system—veins, arteries, capillaries—as a balloon, the wall of the balloon provides resistance to its expansion. Now think of the blood that occupies the vascular system as water. So we have that weapon of the adolescent—the water balloon.

Now ask what controls the pressure inside the balloon, our model of the human vascular system. There are two points. First, anything that tends to add water to the balloon will increase the pressure; anything that diminishes the amount of water in the balloon will decrease the pressure. Second, anything that increases the resistance to expansion of the balloon will increase the pressure and vice versa. That is basically all we need to know to understand how the R-A-A system controls blood pressure.

Here is one simple example to illustrate the utility of this simple model of the human vascular system. As noted earlier, diuretics are agents that act on the kidneys to increase the excretion of water from the human body. In our model, this corresponds to decreasing the amount of water in the balloon. The loss of water should decrease the pressure, as indeed it does, as Karl Beyer demonstrated many years ago.

The essential features of the R-A-A system are presented in Figure 7.1. Let's see how this works. We begin at the top of Figure 7.1 with angiotensinogen. I understand that this is a difficult and meaningless word to many, but that is its name and so I have no choice but to use it. Bear with me.

Human angiotensinogen is a protein that circulates in the blood. It consists of a chain of 452 amino acids linked together. A space-filling model of angiotensinogen was provided in Figure 4.5. Angiotensinogen has no known function in human physiology except to serve as a precursor for small peptides (short chains of amino acids) that do have something to do. Here is what happens.

The enzyme renin, which we met back in chapter 3, synthesized and released into the blood by the kidneys, catalyzes the cleavage of the first 10 amino acids of the angiotensinogen molecule to form a new peptide—angiotensin I (A-I)—plus the rest of the angiotensinogen molecule, about which I shall say no more.

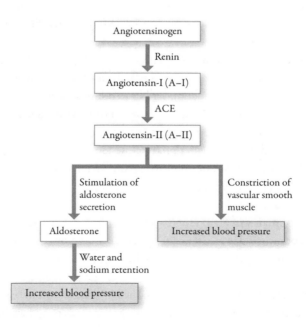

FIGURE 7.1 The Renin–Angiotensin–Aldosterone System.

$$^1\text{Asp-Arg-Val-Tyr-Ile-His-Pro-Phe-His-Leu}^{10}$$

$$^1\text{Asp-Arg-Val-Tyr-Ile-His-Pro-Phe}^8 + \text{His-Leu}$$

Angiotensin-II (A-II)

FIGURE 7.2 Formation of Angiotensin II. Angiotensin converting enzyme, ACE, catalyzes the cleavage of the His–Leu dipeptide from the end of angiotensin I to create angiotensin II, a molecule critically involved in the control of blood pressure.

A-I, too, has little physiological role, but it gets us closer to the action. It serves as the precursor for a biologically active peptide angiotensin II (A-II), the formation of which is catalyzed by ACE. ACE chops the last two amino acids off the end of A-I (Figure 7.2):

The three-letter abbreviations in Figure 7.2 stand for individual amino acids. The only thing worth remembering here is that ACE chops off the last two amino acids of A-I to generate A-II. This will be of interest later.

Now we are getting someplace because A-II has two powerful physiological actions that affect blood pressure (see Figure 7.1). Also, it may now be clearer why the enzyme doing the work is called *angiotensin-converting enzyme*. It catalyzes the conversion of A-I to A-II.

First, A-II causes contraction of the smooth muscle cells that line the walls of the blood vessels. An illustration of the structure of a human artery is provided in Figure 7.3. In terms of our water balloon model, this contraction increases pressure in the balloon. Anything that tends to diminish the volume of the vasculature increases the pressure within it.

Second, A-II stimulates the secretion of the steroid hormone aldosterone from the adrenal glands that ride on top of the kidneys. Aldosterone acts to increase the retention of water and sodium in the body. Thus, aldosterone action is equivalent to adding water to our water balloon, increasing the pressure inside.

In sum, the actions of A-II tend to increase blood pressure by constricting vascular smooth muscle and increasing the volume of fluid in the vascular system through the actions of aldosterone. Given these facts, it seems reasonable that a drug that inhibits the synthesis of A-II would reduce blood pressure. Because ACE is the enzyme that catalyzes the synthesis of A-II, it is the target for inhibitor discovery. If ACE is inhibited from doing its work, then we would expect A-I to pile up and A-II to diminish (Figure 7.4). These changes should permit the vascular smooth muscle cells to relax and let the kidneys get rid of some water, and blood pressure should decrease. And this is what happens.

Before we get any further into our story, let's recognize something. The R-A-A system exists for a reason other than to give the pharma industry targets for drug discovery.

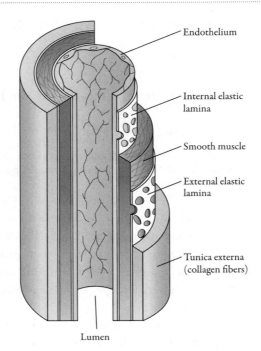

Endothelium

Internal elastic
lamina

Smooth muscle

External elastic
lamina

Tunica externa
(collagen fibers)

Lumen

FIGURE 7.3 Structure of a Human Artery. The key structure is the layer of smooth muscles cells enclosed between the internal and external elastic lamina. This is the site of action of angiotensin II, which acts to constrict this smooth muscle cell layer, increasing the resistance of the artery to expansion and increasing blood pressure.

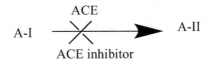

ACE

A-I ⟶ A-II

ACE inhibitor

FIGURE 7.4 The Action of an ACE Inhibitor. The point here is that an effective inhibitor of the action of ACE decreases or prevents the conversion of angiotensin I to angiotensin II. The net effect will be a reduction in blood pressure.

The R-A-A system is one of the body's several ways of controlling blood pressure. Here is how it works to protect us from an unsafe drop in blood pressure (it can be too high, but it can also be too low). In response to events, extensive loss of blood for example, that lead to a decrease in blood pressure or inadequate blood flow through the kidneys, increased quantities of renin are released from special cells in the kidneys. This increase in renin activity leads to increases in the levels of A-I and, subsequently through the action of ACE, A-II. The actions of A-II to constrict blood vessels and increase blood volume tend to restore blood pressure to safe levels. After this happens, renin release from the kidneys decreases. Hence, the R-A-A system is a negative feedback regulatory system; have a look at Figure 7.5.

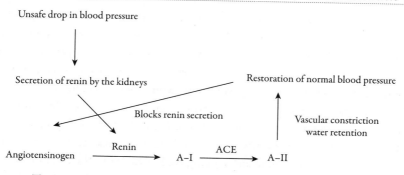

FIGURE 7.5 The R-A-A System: A Negative Feedback Regulatory System. The R-A-A system completes a negative feedback loop. The secretion of renin by the kidneys as a result of an unsafe decrease in blood pressure restores normal blood pressure through the actions of renin and ACE. After this happens, renin secretion is inhibited and the system returns to a state of balance. Negative feedback loops are self-regulating and are inherently safe.

Squibb Leaped First

It happens that, at the time, only Squibb focused significant resources on discovering inhibitors of ACE. Why should this have been so? There are several reasons why others hesitated or ignored the opportunity. About 5 percent of people with high blood pressure have abnormally high levels of circulating renin. These people are called *high-renin hypertensives*. The conventional wisdom of the time was that an ACE inhibitor would work to lower blood pressure only in patients with high renin. Most pharma houses were not excited about devoting resources—and it takes a lot of resources to get a blood pressure-lowering agent on the market—to such a modest fraction of the potential population. The conventional wisdom was wrong, as it sometimes is, and the Squibb scientists proved that. Once eyes were opened, everybody joined in the game.

There were other reasons to be wary of ACE inhibitors. Let's compare ACE with renin. Renin could be a better target because it is a very specific enzyme; it clips 10 amino acids off the front of angiotensinogen to liberate A-I but does nothing else. Hence, it is reasonably possible to predict the physiological consequences of a renin inhibitor. This is not so for an ACE inhibitor. ACE is less particular about what it acts on than renin. For example, the breakdown of the peptide bradykinin, which relaxes arterial smooth muscle cells, is also catalyzed by ACE. Thus, ACE inhibition might well increase bradykinin levels, which complicates the picture. Elevated bradykinin might help lower blood pressure, but may also contribute to unwanted adverse effects of its own.

Furthermore, it was not clear that ACE action was the only route to A-II. Other enzymes were known to be able to convert A-I to A-II. Thus, they might generate A-II even though ACE was inhibited. Whether they would do so to a significant extent in the human body was not clear. These issues are adequate to illustrate the complexities that scientists face when they try to choose a molecular target for drug discovery work.

Despite these concerns, Squibb did move ahead and establish the efficacy and safety of ACE inhibitors for control of high blood pressure and for other indications, as well. The heroes of the Squibb story are Miguel Ondetti, a chemist, and David Cushman, a biologist. Both have been abundantly, but perhaps not adequately, honored for their groundbreaking contributions to human health. More about them follows a bit later.

I was at Merck, heading their biochemistry program, at the time that Squibb established that ACE inhibitors were a hot property. Merck had had a very small effort directed at ACE—just two chemists, as I recall. Within a week, 20 chemists had been assigned to the task of designing and making ACE inhibitors. The chemists were supported by a comparable number of biologists. This was the beginning, and the effort grew and paid off. Under the leadership of Art Patchett, Merck introduced the next two ACE inhibitors into clinical practice.

I am going to get around to how the Squibb scientists, and subsequently the Merck scientists, discovered ACE inhibitors. It is an interesting story based, in significant part, on the chemistry of snake venoms. Before we get there, it is worthwhile to have one last look at the R-A-A system. It reveals other useful ways to control blood pressure.

Beyond ACE inhibitors, the R-A-A system suggests three ways to get at blood pressure control. Starting again at the top of Figure 7.1, consider inhibition of the first enzyme in the pathway—renin. A renin inhibitor should reduce the level of A-I, and without A-I, no A-II can be generated by the action of ACE. So inhibiting renin is a reasonable alternative to inhibiting ACE. It is not only reasonable but it works. The renin inhibitor, aliskerin, marketed as Tekturna, was recently introduced into clinical practice for control of high blood pressure, just as the ACE inhibitors. In passing, let me note that the detailed structure of renin was invaluable in the design of aliskerin. It is the key that fits the renin lock—the groove visible at the top of the structures shown in chapter 3. It fits the lock but does not open it. It is a renin inhibitor.

We can also imagine drugs that would antagonize the *action*, rather than the *synthesis*, of A-II. It makes little difference if there is lots of A-II around if it cannot act to constrict vascular smooth muscle or induce synthesis of aldosterone. Drugs in this class are known as *angiotensin receptor blockers* (ARBs). A-II works through receptor molecules. These are proteins, but not enzymes. When A-II links up with its receptors, a series of consequences ensue leading to effects on vascular smooth muscle and aldosterone synthesis. Finding an antagonist at a receptor is a lot like finding an inhibitor of an enzyme. There is a family of ARBs approved for control of blood pressure, heart failure, and kidney protection in diabetics. The first to be approved was losartan, marketed as Cozaar. Several more have followed. ARBs are an important class of drugs in cardiovascular medicine.

Last, we can envision antagonists of aldosterone. Drugs in this class would blunt the effect of this water- and sodium-retaining hormone. Here, too, we have drugs in this class that are useful for blood pressure control. Two examples are spironolactone (Aldactone) and eplerenone (Inspra). Targets in the R-A-A system have been exploited systematically to discover and develop these four classes of useful drugs: ACE inhibitors, a renin inhibitor, ARBs, and aldosterone antagonists.

The Chemistry of Snake Venoms Showed the Way

After Squibb scientists, under the leadership of Ondetti and Cushman, had elected to explore the utilities of ACE inhibitors in human medicine, they had an obvious problem: how does one find such inhibitors? The detailed structure of ACE was not known, so trying to design a key to fit the ACE active site, the lock, was not an option. If you do not know the structure of the lock, you cannot design a key. Natural product chemistry came to the rescue.

Living organisms create a wealth of interesting molecules. The chemistry of these molecules is known as *natural products chemistry*. Natural product chemists isolate these molecules, determine their structure, synthesize them, and explore their biological properties. Such searches have turned up riches. About 40 percent of all drugs used in human medicine are natural products or are derived from them. Chemists are an inquisitive lot and look for natural products in places that might not quickly occur to you. One of these areas is snake venoms. Peptides isolated from snake venoms provided the first potent inhibitors of ACE, and an opportunity to explore the pharmacological and toxicological consequences of ACE inhibition.

A Brazilian scientist, Sergio Ferreira, discovered that the venom of a South American pit viper, *Bothrops jararaca*, contained peptides capable of potentiating the action of bradykinin. Recall that bradykinin is a smooth muscle relaxing agent and a substrate for ACE that inactivates it. So Ferreira's observation strongly suggested, but did not prove, that these peptides are ACE inhibitors. Ferreira isolated nine peptides from his crude mixture, determined their structure, and synthesized one of them known as *BPP5a*. Ferreira then joined the laboratory of noted American pharmacologist (and subsequent Nobel Laureate) John Vane as a postdoctoral fellow, bringing his snake venom peptides with him. Vane was able to show that BPP5a inhibited the conversion of A-I to A-II in living animals, establishing that BPP5a is an effective inhibitor of ACE. Independently, Ondetti and his coworkers fractionated the crude snake venom, isolated six peptides, determined their structure, and synthesized them. The most potent of the lot was a peptide nine amino acids long named *teprotide*.

At the same time, Japanese chemists Hisao Kato and Tomoji Suzuki isolated five bradykinin-potentiating peptides from the venom of the poisonous Chinese pit viper *Agkistrodon halys blomhoffii*. Following our earlier logic, these peptides were presumably ACE inhibitors. Kato and Suzuki determined the structures and demonstrated that they were, in fact, inhibitors of ACE. ACE inhibitor peptides from both snake venoms terminated in an amino acid named *proline*. This finding provided an important clue to finding a structure for a clinically useful drug, as we shall see.

Safety assessment studies established that teprotide, isolated by Ondetti, was acceptable for clinical studies. These studies need to be seen as part of drug discovery, hypothesis testing, not an effort to develop a marketable product. Teprotide is not orally bioavailable. That is to say, if taken orally, teprotide does not enter the systemic circulation; it goes in one end and out the other. A drug for routine control of high blood pressure must be taken orally because it needs to be taken daily for a lifetime. So teprotide does not offer the convenience to be a drug. It was, however, a great research tool to test the hypothesis: would an ACE inhibitor prove safe and effective for control of high blood pressure? In clinical studies designed to answer this question, teprotide was given by injection.

Teprotide proved to lower blood pressure effectively in a variety of clinical settings characterized by normal as well as elevated renin levels, and augmented the blood pressure-lowering effects of diuretics. In addition, its effects on the characteristics of blood flow (hemodynamics) predicted the utility of ACE inhibitors for the treatment of heart failure. Adverse effects of teprotide were minimal. This work confirmed the hypothesis that ACE inhibitors would be safe and effective for control of blood pressure.

In summary, work with the ACE inhibitor peptides of snake venoms proved critical to the discovery and development of ACE inhibitors in two senses. First, they provided proof of concept for the efficacy of ACE inhibitors for blood pressure control. Second, they provided strong, useful clues to the structures of potent ACE inhibitors.

Now the issue was how to discover ACE inhibitors with good oral activity and the potency to inhibit ACE and to lower blood pressure that were at least as good as that of teprotide. The structures of teprotide and BPP5a provided important clues, including the fact that both contained a terminal proline amino acid, which brings us to the next part of the story.

Studies of the Enzyme Carboxypeptidase A Suggest a Solution

ACE was a difficult enzyme to target. It is a large, zinc-containing enzyme of unknown three-dimensional structure. It offered few clues for effective small-molecule inhibitor design. In contrast, the enzyme carboxypeptidase A is small, was well characterized at the time, and had a known three-dimensional structure thanks to the crystallographic work of Harvard chemist (and future Nobel Laureate) William Lipscomb. Carboxypeptidase A is uninteresting as a target for drug discovery. It is an enzyme secreted into the gut from the pancreas to aid in the digestion of proteins. Biochemists had studied carboxypeptidase A intensively to gain insights into how enzymes work. They chose it because it was easy to access, easy to assay, and had a known three-dimensional structure.

In an impressive leap of imagination, Ondetti and Cushman reasoned that studies with carboxypeptidase A might be revealing for the design of ACE inhibitors. To be more specific, the two enzymes have two things in common. First, each contains exactly one atom of zinc per protein molecule, and that atom is crucial for enzyme activity. So, there is a key commonality in the catalytic process. Second, carboxypeptidase A lops off one amino acid from the end of peptide chains. ACE does the same, but lops off two at a time. So both enzymes do basically the same thing, although ACE takes a bigger bite (Figure 7.6).

Two biochemists at the University of North Carolina, Richard Wolfenden and his student Larry Byers, were exploring a novel idea for the design of enzyme inhibitors and had chosen carboxypeptidase A as their target. Here is the idea.

Just as enzymes bind their substrates at the active site, they bind their products. When carboxypeptidase A (or ACE) cleaves a peptide chain, it creates two products; if you cut a string in two, you get two strings. Both products must have binding sites on the enzyme. Wolfenden and Byers reasoned that designing a molecule that shared structural elements with both products might bind tightly to the enzyme. Figure 7.7 captures the basic idea. Wolfenden and Byers synthesized a few molecules based on this model and tested them. They proved effective inhibitors of carboxypeptidase A. The structures provided in Figure 7.8 reduce the basic idea to chemistry in the case of the inhibitor L-benzylsuccinic

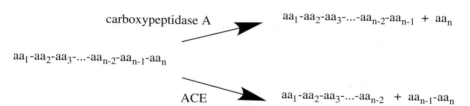

$$\text{carboxypeptidase A} \longrightarrow aa_1\text{-}aa_2\text{-}aa_3\text{-}...\text{-}aa_{n-2}\text{-}aa_{n-1} + aa_n$$

$$aa_1\text{-}aa_2\text{-}aa_3\text{-}...\text{-}aa_{n-2}\text{-}aa_{n-1}\text{-}aa_n$$

$$\text{ACE} \longrightarrow aa_1\text{-}aa_2\text{-}aa_3\text{-}...\text{-}aa_{n-2} + aa_{n-1}\text{-}aa_n$$

FIGURE 7.6 Comparison of the Actions of Carboxypeptidase A and ACE. As shown, carboxypeptidase A liberates one amino acid from the end of an amino acid chain whereas ACE liberates two. Other than this difference, there are compelling similarities in the action of the two enzymes, an important clue to ACE inhibitor design.

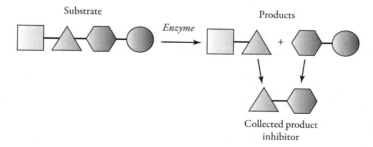

FIGURE 7.7 Collected Product Inhibitor Design. Let's assume that an enzyme cleaves a substrate with four distinct structural elements, symbolized by the four shapes, into two products as shown. The enzyme will bind both product molecules. If we make a hybrid molecule from the tail of one product and the head of the other, we might expect the enzyme to bind that molecule as well, perhaps quite tightly because it has structural elements of both products.

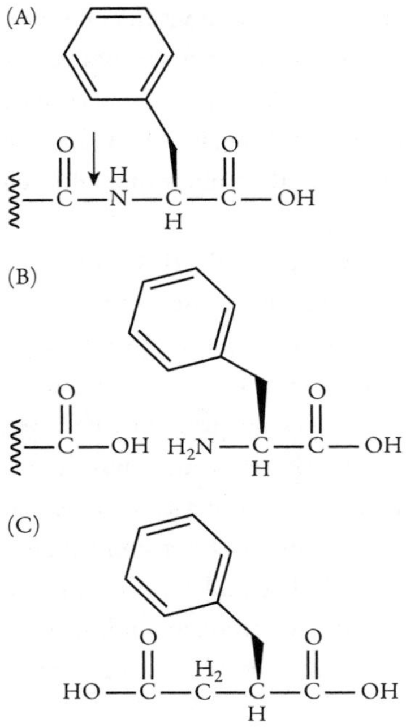

FIGURE 7.8 The Concept of Collected Product Inhibitors in the Case of Carboxypeptidase A. (A) The terminal part of a typical substrate for carboxypeptidase A. The enzyme cleaves the substrate where indicated by the arrow. (B) These are the products of substrate cleavage by carboxypeptidase A. Only the terminal part of one product is shown. (C) This view shows L-benzylsuccinic acid, a molecule that has elements of the structure of both products. It is an inhibitor of carboxypeptidase A. The -COOH group on the left binds to the zinc atom in the enzyme.

acid. Have a look at Figure 7.9 for an up-close look at this inhibitor bound to the active site of carboxypeptidase A.

The collected product inhibitor concept provided the final piece of the puzzle for Ondetti and Cushman. They realized that the carboxyl group (-COOH) of L-benzyl-succinic acid bound to the catalytically crucial zinc atom of carboxypeptidase A, as shown in Figure 7.9. They replaced this carboxyl group with a sulfhydryl group (-SH) that binds better to zinc (zinc loves sulfur). Combining the -SH function with structural units from the terminus of BPP5a and teprotide provided captopril, a potent, orally active, long-acting ACE inhibitor (Figure 7.10). The rest, as they say, is history.

The importance of basic research in enabling this therapeutic breakthrough must not be overlooked. The natural products chemistry focused on snake venoms. The proof of the underlying rationale with two of them, the structural and mechanistic studies of carboxypeptidase A, and the discovery of a novel inhibitor design principle in an academic lab are rungs on this ladder of success. Full credit goes to Ondetti and Cushman for having the wit and determination to access all that was available and pull it together to create captopril. As is so often the case, the concepts and work of many have provided the background required for successful innovation.

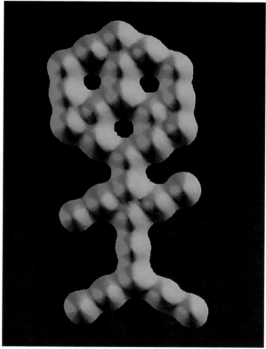

FIGURE 2.3 Individual Atoms and Molecules Can Be Manipulated. (A) Assembling atoms on a metal surface creates the Japanese characters for the word *atom*. (B) Molecules of carbon monoxide, CO, create a cartoon of a man. Both images were generated by the technique of scanning tunneling microscopy and are provided courtesy of IBM.

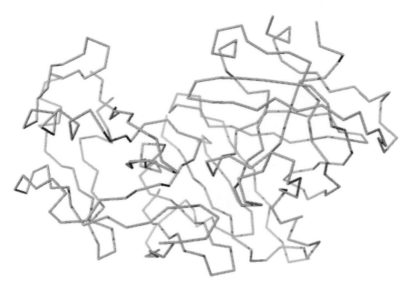

FIGURE 3.5 The Folding of the Main Chain of Human Renin. This figure presents the main chain of human renin, an enzyme secreted by the kidney and involved in the control of blood pressure. No individual atoms are shown. The point of this figure is to provide an example of the complexity with which the amino acid chains of proteins fold up in space to create a compact structure. You can identify one end of the chain in the upper left of this figure and the other end at the bottom center. If you are careful, you can trace the chain as it weaves back and forth, up and down, to create the compact structure of the active enzyme. Note that the darker lines should be thought of as closer to you and the lighter ones as farther from you.

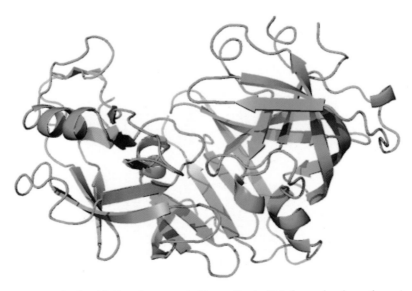

FIGURE 3.6 α Helical and β Sheet Structures in Human Renin. This figure also shows the main chain of human renin, but with some additional information. Some parts of the amino acid chain are organized into helices, termed the α helix, and these are shown as ribbons. Other parts of the amino acid chain loop back on themselves and form sheet structures, termed β sheets. These are shown as thick arrows. The remainder of the chain occupies specific spots in space, but is not organized into helices or chains, shown here in the narrow lines.

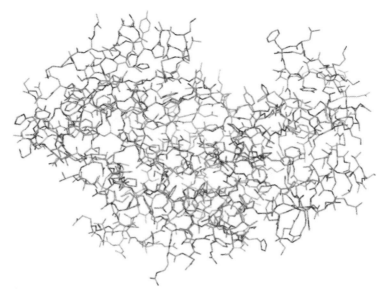

FIGURE 3.7 A Detailed Structure for Human Renin. This figure is a line representation of the structure of human renin, revealed in a more detailed way than in Figures 3.5 and 3.6. Here, chemical bonds linking atoms are shown as lines. Every place where two lines intersect or terminate is the site of an atom. Carbon atoms are shown in green, oxygen atoms in red, nitrogen atoms in blue, and sulfur atoms in gold (these are difficult to find because there are only a few of them). To keep the structure reasonably simple, hydrogen atoms are not shown. Note the compact structure formed. The cleft toward the top of the figure is the site where renin interacts with its substrates and where catalysis takes place. Note that this cleft is also evident in Figures 3.5 and 3.6.

FIGURE 3.8 A Space-Filling Model of Human Renin. Here in this space-filling model of human renin, individual atoms are shown as spheres. The point of this representation is to show how compact the structure is. Human renin wraps up quite tightly so that there is little unoccupied room in the molecular interior. Carbon atoms are shown in green, oxygen atoms in red, nitrogen atoms in blue, and sulfur atoms in gold (you can identify a few at the top of the structure).

FIGURE 4.4 Space-Filling Model of the Protein Human Angiotensinogen, the Physiological Substrate of Renin (Source: Zhou, A., R. W. Carrell, M. P. Murphy, Z. Wei, Y. Yan, P. L. Stanley, P. E. Stein, F. Broughton Pipkin, and R. J. Read. 2010. A redox switch in angiotensinogen modulates angiotensin release. Nature 468: 108–111.)

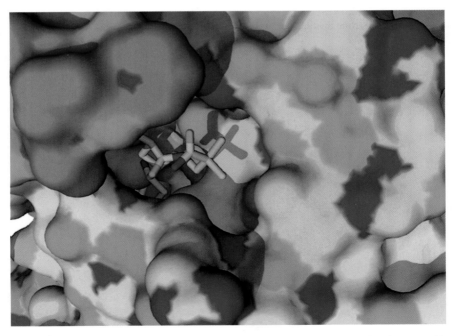

FIGURE 4.5 A Model of a Small-Molecule Inhibitor of Renin—Aliskerin—Bound to the Active Site of the Enzyme. The surface contours of the atoms at the active site of renin are shown in orange. Aliskerin is shown as a stick figure in which each bend in the molecule represents an atom. (Source: Rahuel, J., V. Rasetti, J. Maibaum, H. Reuger, R. Goschke, N. C. Cohen, S. Stutz, F. Cumin, W. Fuhrer, J. M. Wood, and M. G. Grutter. 2000. Structure-based drug design: The discovery of novel, nonpeptide, orally active inhibitors of human renin. *Chem Biol* 7: 493–504.)

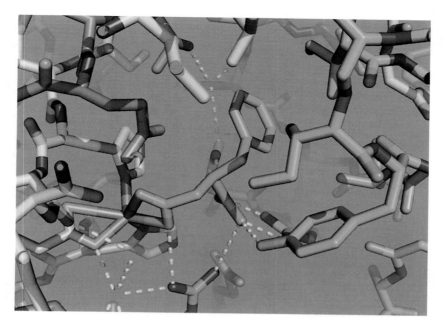

FIGURE 7.9 A Portion of the Active Site of Bovine Carboxypeptidase A Complexed with L-Benzylsuccinic Acid. This is one of the inhibitors designed by Wolfenden and Byers that provided insight into inhibitor design for ACE. The structure of L-benzylsuccinic acid is shown in Figure 7.8. Here, the carbon atoms of the inhibitor are shown in green and the oxygen atoms in red. One of the COOH groups is shown bound to the single zinc atom of the enzyme, shown here in gray (left side of the image). The other COOH group (pointing down in the center of the image) interacts with amino acid side chains through hydrogen bonds (dashed yellow lines). The side chain of the inhibitor occupies a hydrophobic pocket in the enzyme (right side of the image).

FIGURE 8.9 Space-Filling Model of Hydroxymethylglutaryl-CoA Reductase (HMGR). HMGR is the molecular target of the statins. There is a single chain of 888 amino acids that assembles into a quite compact three-dimensional structure.

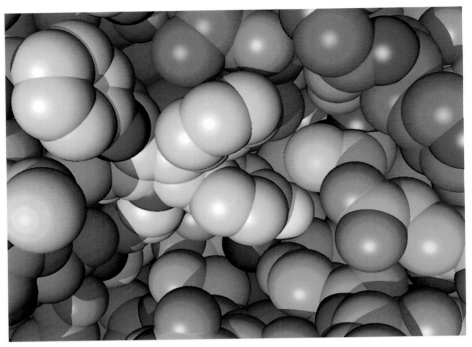

FIGURE 8.10 Close-up of a Statin—Atorvastatin (Lipitor)—in the Active Site of HMGR. Atorvastatin in blue is nestled into the active site of HMGR shown in a space-filling model.

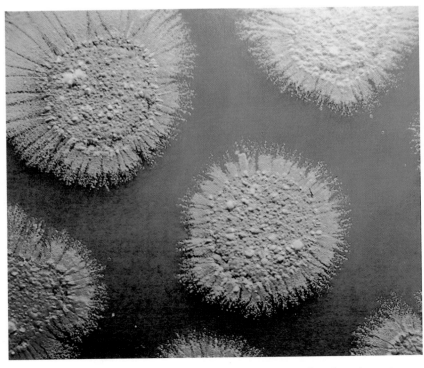

FIGURE 9.3 *Streptomyces cattleya*. This photograph of *Streptomyces cattleya* shows it growing as colonies on a large plate. Note the color of these colonies, reminiscent of that of cattleya orchids, from which its name is derived.

FIGURE 10.4 In Wau, South Sudan, Joseph Bringi Racardo and his wife, Rosa, are both blind from onchocerciasis. A child from their village has offered to assist them on their way. Around the world, onchocerciasis has an enormous impact, preventing people from working, harvesting crops, receiving an education, or taking care of children. (Courtesy of the Carter Center/E. Staub.)

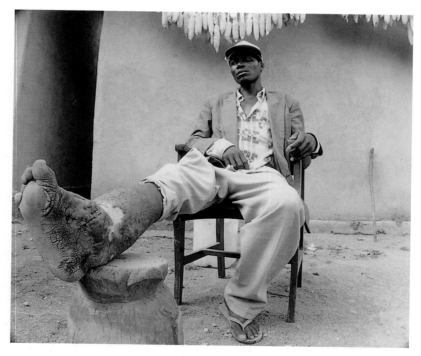

FIGURE 10.5 In Nigeria, Joseph Gondovo, a victim of lymphatic filariasis, sits with his foot elevated. Gondovo had suffered from a swollen leg for years, but after joining a lymphatic filariasis support group led by The Carter Center, he has learned techniques, including better diet and hygiene, that have improved his condition. (Courtesy of the Carter Center/E. Staub.)

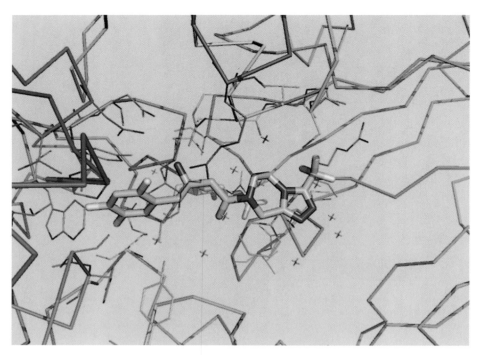

FIGURE 12.11 Sitagliptin Occupying the Active Cite of Dipeptidyl Peptidase 4 (DPP-4). Sitagliptin is shown as a stick figure, mostly blue. The surrounding DPP-4 structure is shown as simple line segments.

FIGURE 12.14 Ann E. Weber and Nancy A. Thornberry, winners of the 2011 Discoverers Prize from the Pharmaceutical Research and Manufacturers of America.

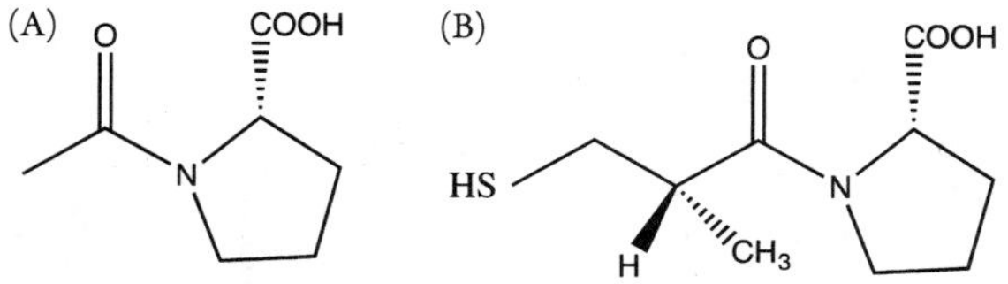

FIGURE 7.10 (A) Terminal structure of teprotide and BPP5a. (B) Captopril. Note the similarity in structure to the terminus of teprotide and BPP5a.

Merck Elects to Compete

As I noted earlier in this book, the first product to market in a particular class is not always either the best one or the most successful in the marketplace. Consider the statins, to which I turn in chapter 8. The first to market was Mevacor, but the biggest commercial success was Lipitor, the sixth statin to be launched. The prospect of perhaps creating a superior product coupled with that of reaping some commercial gain is usually compelling in the pharmaceutical industry. Follow-the-leader is frequently a good strategy, as many successful companies will testify. Perhaps the most compelling factor is that the risk associated with selection of the target has been removed. The pioneer has taken that risk out of the equation.

So Merck leaped into the ACE inhibitor competition and named chemist Art Patchett to lead the charge. I knew Art quite well. In my mind, Art was the most creative medicinal chemist at Merck—and Merck had a great group of medicinal chemists.

There was a rationale that Merck adopted in fashioning its drug design goals. The clinical development of captopril revealed a safety profile that, although surely acceptable, left room for improvement. The side effect profile was, in some ways, reminiscent of that for penicillamine (including rash and itching). Captopril and penicillamine share a common structural element, a sulfhydryl group, -SH. In other words, a hydrogen atom is linked to a sulfur atom that, in turn, is linked to a carbon atom somewhere in the molecule. As noted earlier, the sulfhydryl group in captopril is critical; it links to the single zinc atom in ACE. This interaction contributes a great deal to the affinity of captopril for ACE. Deleting that structural element from an ACE inhibitor while retaining inhibitor potency was a tall order. Nonetheless, that is what Art and his coworkers needed to achieve to come up with a competitive ACE inhibitor. Playing follow-the-leader is not so easy. You need to get around the patent position of the pioneer in the field and come up with a molecule that is, in one way or another, superior to what is available on the market. Put another way, the marketing department needs to have a "hook" to be successful in the marketplace (although sometimes they succeed without one).

In retrospect, I believe that Merck's rationale was wrong. The initial less-than-optimal safety profile of captopril was probably the result of using too high a dose of the drug

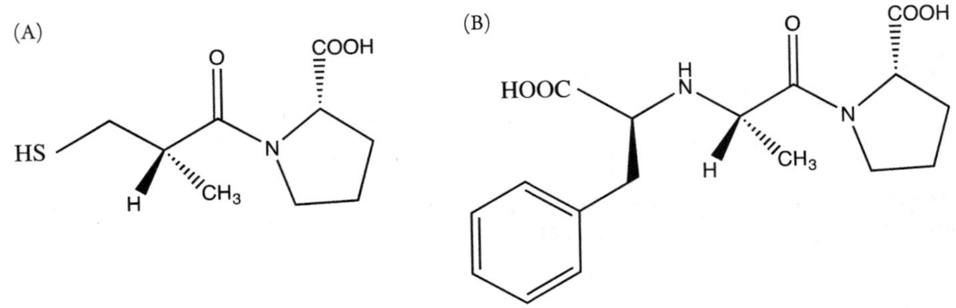

FIGURE 7.11 (A) Captopril. (B) Enalaprilat.

during the clinical trials. When a lower dose was used after the drug was launched, the safety profile improved. Right or not, getting rid of the sulfhydryl group was the goal that drove the Merck design effort.

This was not so easy. The zinc atom has a high affinity for the sulfhydryl group and, in consequence, it added a lot to the potency of captopril. Art and his colleagues needed to find some way to compensate for the loss of potency resulting from deletion of the sulfhydryl group. They worked it out as shown in Figure 7.11.[1]

Enalaprilat shares the structure on the right side of the molecule with captopril (and with teprotide and BPP5a). On the left side of the molecule, the SH group has been replaced by a COOH group that binds to the zinc atom of ACE, as does the SH group, but less tightly. To make up for this loss, enalaprilat includes a bulky, hydrophobic group that occupies a pocket within the active site of ACE that captopril ignores. The energy of this interaction compensates for the loss of affinity for the zinc atom. Captopril and enalapril are comparably potent inhibitors of ACE.

Although enalaprilat is the effective inhibitor of ACE, the marketed product is enalapril, a prodrug of enalaprilat (the ethyl ester, to be specific). Enalapril has better oral bioavailability than enalaprilat and is converted rapidly to enalaprilat in the body.

As always, getting a molecule into clinical trails involves more than chemistry. In the case of enalapril and lisinopril (Merck's second entry in the ACE inhibitor field), Art's chemistry efforts were married to an outstanding pharmacology program headed by Charles Sweet. The work of Patchett and Sweet was aided by that of perhaps 100 to 200 others at Merck—biochemists, physiologists, pharmacologists, toxicologists, pathologists, statisticians, clinicians, and more. The clinical programs in support of these product candidates went smoothly. Merck recognized Art Patchett and Charles Sweet with Director's Scientific Awards.

The result of this effort was, as noted earlier, the second and third ACE inhibitors to enter the marketplace. The FDA approved enalapril, marketed as Vasotec and Renitec, in 1985 and lisinopril, marketed as Zestril and Prinivil, in 1987. These are the hallelujah moments for these drugs. At the end of the day, they proved to be the most successful products among the ACE inhibitors.

Enalapril has been combined with two calcium entry blockers, also used to treat high blood pressure. The combination of enalapril and felodipine is marketed as Lexxel; enalapril and diltiazen are marketed at Texzem.

Other ACE inhibitors have been discovered and developed, but enough for this story. Let's move on to the statins, which are drugs that lower cholesterol levels (and a story in which Art Patchett again played a key role).

8 Statins
PROTECTION AGAINST HEART ATTACKS AND
STROKES—HALLELUJAH!

CHOLESTEROL! THIS MAY be the single most famous (or infamous) small molecule of life. Most people view it as a threat to good health and even to life itself. We search for foods that are cholesterol free or at least low in cholesterol. We use them in efforts to achieve a low-cholesterol diet. Our primary care physicians measure our blood cholesterol levels routinely and report the news, good and bad. If the level is high, they recommend a better diet (that is, one lower in cholesterol and saturated fat), more exercise, and perhaps weight reduction. If those measures fail to get the cholesterol level where it should be, it is highly likely that therapy with a cholesterol-lowering drug will be recommended. The drug will usually fall into a class known as *statins*. Statins are among the most frequently prescribed drugs in the world.[1]

The first statin approved for marketing by the FDA in the United States was lovastatin (Mevacor), which happened in 1987. Lovastatin was followed into clinical practice by pravastatin (Pravachol), simvastatin (Zocor), fluvastatin (Lescol), atorvastatin (Lipitor), cerivastatin (Baychol),[2] pitivastatin (Livalo), and rosuvastatin (Crestor). There are a lot of options from which to choose among the statins.

The story of how statins were discovered and developed is pretty amazing. The tale focuses on cholesterol in its several dimensions—what it is, how it is made, how its levels are regulated, the health consequences that may ensue when proper regulation fails, and how statins act to restore that regulation. The task of this chapter is to tell the tale. The focal point is cholesterol. So that is where we begin.

Cholesterol Is an Essential Molecule of Life

There are two sides to most stories, which is certainly the case for cholesterol. Although what we hear about cholesterol is mostly negative (isn't there some way to get rid of this stuff?), the fact is, we cannot live without it[3] and there are three reasons why.

First, cholesterol is an essential component of all our membranes. Cholesterol, a family of phospholipids, and certain proteins are the building blocks for the membranes that isolate the interior of cells from their surroundings. They are also the building blocks for all the membranes that exist within our cells, including those that isolate the cell nucleus from the cell cytoplasm; those that define the mitochondria, the lysosomes, and the endoplasmic reticulum; and more. The cell is the unit of structure and function in living organisms. Its integrity depends on its membranes.

Second, cholesterol is the essential precursor for all our steroid hormones. Cholesterol is converted into androgens (male sex hormones), estrogens (female sex hormones), glucocorticoids, progesterone, and mineralocorticoids. These hormones play a series of critical roles in human physiology. We cannot get along without them. And without cholesterol, we would not have them.

Third, cholesterol is the precursor for the bile acids. These molecules are made in the liver and dumped into the gut by way of the gall bladder after a meal. Their task is to aid in the digestion of the fats in our diet. Fats are insoluble in water. For the digestive enzymes—known as *lipases*—to degrade them effectively, the fats must be dispersed into very small droplets and form an emulsion, which creates a tremendous surface area on which the lipases can work their digestive action. The role of the bile acids is to act as the emulsifying agents. If you have made hollandaise or béarnaise sauces, you already know the idea. The lecithin molecules in the egg yolks act to emulsify the butter fats and the wine, and something rather good emerges. And something like this is what happens in your gut after a fat-containing meal. Conversion to bile acids is the largest single drain on the cholesterol made in the liver. Perhaps we could live without bile acids, but life is surely better with them.

So the issue with cholesterol is not whether it is important; it surely is. The problem is when blood levels exceed established levels for good health. Elevated blood cholesterol can lead to atherosclerosis and its sequelae: heart attacks, strokes, and peripheral artery disease. What can we do about this?

Getting a good answer to this question depended on a lot of basic research. As I emphasized earlier, advances in health care almost always depend on basic research findings, research undertaken out of curiosity about how things work and why. Matters of great utility and practicality derive from rather arcane beginnings. This may be hard to understand, but it is quite true nonetheless. So let's have a quick look at what we had to learn to discover the statins.

The Biosynthesis of Cholesterol Is Complex

We get cholesterol from two sources: our diet and what our body makes, primarily in the liver. To a significant degree, these two sources act in concert. If you restrict your dietary intake of cholesterol and saturated fat, the liver makes more. If your diet is rich in cholesterol and saturated fat, the liver makes less. They do not compensate completely. If you control your diet rigorously, you can reduce your plasma cholesterol. *Rigorously* is the operative word here. Because most of us with high cholesterol levels do not succeed in reaching the right level through lifestyle changes, we need help. The help must come from regulating the dynamics of the cholesterol that we make. So we need to understand those dynamics.

Cholesterol is a relatively small but complex molecule. It contains a skeleton of four rings made from carbon atoms fused together (Figure 8.1).

As you can see in Figure 8.1, the skeleton is decorated with a number of atomic groupings. The structure of cholesterol was deduced during the 1930s in a heroic chemical undertaking. The question then became: how is it made in the body?

The structure of cholesterol offers no clues about how it is assembled. Unraveling that secret of nature required 20 years of inspired science. Unraveling the secret of how the assembly is regulated required several more. Along the way, 13 Nobel Prizes in either Chemistry, or Physiology or Medicine were awarded to scientists who spent a significant fraction of their careers studying cholesterol.

There were several keys to mapping the biosynthetic route to cholesterol and I shall not delve into all of them. A few of them are worth our time. Perhaps the most important technology used was that of isotopic tracers. Elements come in two or more flavors. The number of protons in the atomic nucleus defines each element. For example, carbon has six protons. The various flavors derive from variability in the number of neutrons in

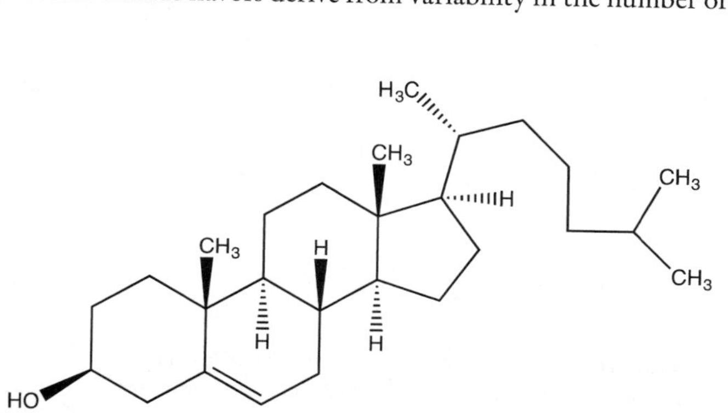

FIGURE 8.1 Cholesterol.

the atomic nucleus. About 98.9 percent of all carbon atoms have six neutrons to neigh-bor with the six protons in the nucleus of a carbon atom. Most of the rest have seven neutrons in the atomic nucleus. A very small fraction of carbon atoms have eight nuclear neutrons; such atoms are known as *carbon-14* or ^{14}C. The point here is that carbon-14 is radioactive. The radioactivity serves as a label that can be detected with, for example, a Geiger counter. Suppose we wanted to know whether the simple two-carbon molecule acetic acid—the stuff in your balsamic vinegar—is a precursor of cholesterol. You could obtain acetic acid containing carbon-14 in one or the other of its carbon atoms, add it to a biological system that makes cholesterol, and determine whether any radioactivity shows up in the cholesterol.

Konrad Bloch, a biochemist then at the University of Chicago, later at Harvard, did that experiment in 1952 and demonstrated that several carbon atoms of cholesterol are, in fact, derived from acetic acid without dilution of the radioactive label. So acetic acid was converted into cholesterol. Bloch further hypothesized that all the 27 carbon atoms of cholesterol were derived from acetic acid.

John Cornforth, an Australian chemist, and George Popják, a Hungarian chemist, working together in England proved Bloch correct. They devised a very clever chemical scheme to dissect the cholesterol molecule atom by atom. Here is the structure of acetic acid: CH_3-COOH. The CH_3 group of atoms is known as a *methyl group*; the COOH group of atoms in known as a *carboxylic acid group*. This is all the chemistry we need to know.

Using acetic acid radiolabeled in one of its carbon atoms—say, the methyl carbon atom—Cornforth and Popják were able to identify each of the atoms of cholesterol derived from that atom of acetic acid. Repeating the entire process with acetic acid radiolabeled in the carboxylic acid carbon, they proved that all the other carbon atoms of cholesterol came from this source. Figure 8.2 shows the result; *m* identifies the methyl group carbon and *c* identifies the carboxylic acid carbon atom of acetic acid.

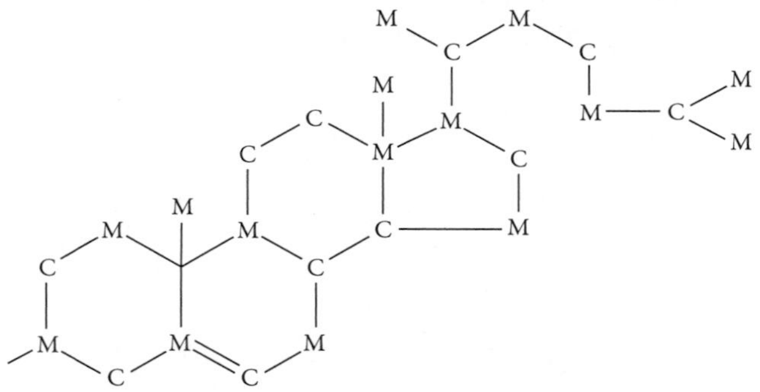

FIGURE 8.2 Sources of the Carbon Atoms of Cholesterol. Here, each carbon atom of cholesterol is labeled either by *m* or *c*. Those labeled *m* are derived from the methyl group of acetic acid. Those labeled *c* are derived from the carboxylic acid carbon of acetic acid. Note that all 27 carbon atoms of this complex molecule are derived from acetic acid, which contains just two carbon atoms.

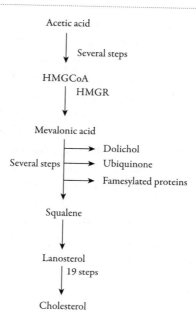

FIGURE 8.3 Summary of the Biosynthetic Route from Acetic Acid to Cholesterol. There are 30 steps along the route from acetic acid to cholesterol. Of particular note is the reaction leading to mevalonic acid catalyzed by hydroxymethylglutaryl-CoA reductase (HMGR). This enzyme is the molecular target of the statins. Squalene is a 30-carbon molecule that is cyclized to create the four-ring skeleton of the steroids. The immediate product is the steroid lanosterol, which is converted to cholesterol in 19 additional steps. Note that the route from mevalonic acid to squalene has branches leading to other molecules of importance: dolichol, ubiquinone, and farnesylated proteins.

Cornforth and Popják completed this amazing task in 1957, and the former went on to win the Nobel Prize in Chemistry in 1975, based in part on this work. So now the task was to figure out the series of chemical steps—the biosynthetic route—by which the simple molecule acetic acid is converted into cholesterol.

A big clue to the biosynthetic route was discovered almost by chance. In 1956, Karl Folkers and his group of chemists at Merck isolated a substance that would replace acetate—acetate-replacing factor—required for growth by a mutant of *Lactobacillus acidophilus* from residues from the fermentation of grains for production of alcoholic beverages. Shortly thereafter, they determined its structure. They named it *mevalonic acid*. Note that *Lactobacillus* does not make cholesterol. Nonetheless, the structure of mevalonic acid was suggestive of a possible role as an intermediate on the way to cholesterol. Merck scientists demonstrated quickly that mevalonic acid is converted to cholesterol more actively than acetic acid in a rat liver preparation. So, presumably, acetic acid is converted to mevalonic acid, which is in turn converted to cholesterol. Of great interest for our story is the discovery of the enzyme that catalyzes the formation of mevalonic acid. Feodor Lynen, a German biochemist, did that in 1957.[4] The formal name of that enzyme is hydroxymethylglutaryl-CoA reductase (CoA derives from Coenzyme A), which we will refer to as *HMGR*. It is the molecular target of the statins.

To cut a very long story short, the route from acetic acid to cholesterol requires 30 chemical steps. Konrad Bloch elucidated much of that route throughout a 20-year period. We know every step in the pathway and have an understanding of every enzyme along the way. Incidentally, we know every step from cholesterol to the steroid hormones and from cholesterol to the bile acids. We also understand a number of pathways to other products that branch off from the route to cholesterol. The general outline of all this work is summarized in Figure 8.3. For his work on the biosynthesis of cholesterol, Bloch was awarded the Nobel Prize for Physiology or Medicine in 1964, which he shared with Feodor Lynen.

Of particular interest for our story is the fact that the reaction leading to mevalonic acid, catalyzed by HMGR, is the rate-determining step for cholesterol biosynthesis. In other words, it is the slowest step along the multistep pathway. Slowing this step slows the overall rate of cholesterol synthesis.

Now that we know something about how cholesterol gets made, the next step in our tale is to understand how its synthesis is regulated.

The Low-Density Lipoprotein Receptor Is a Key Player in the Regulation of Cholesterol Metabolism

Two biomedical scientists at the Southwestern Medical School in Dallas, Texas— Michael Brown and Joseph Goldstein—pretty much figured out how cholesterol metabolism is regulated. They shared the Nobel Prize in Physiology or Medicine in 1985 for their work.

Brown and Goldstein got started through the study of a disease known as *familial hypercholesterolemia* (FH). As the name implies, this is a genetic disease, one that is inherited from one's parents. Also, as the name implies, patients with FH have too much cholesterol in their blood, which is not a good thing. The genetic defect causing FH is in the gene encoding the receptor for low-density lipoproteins (LDLs), often referred to as *bad cholesterol*. Brown and Goldstein discovered the LDL receptor, and it is the star of this part of the story.

Here is the critical experiment that led Brown and Goldstein to the LDL receptor. Human skin fibroblasts in cell culture make cholesterol. When the cells are grown in a simple protein-free medium, the rate of cholesterol synthesis is high. When human serum (blood from which red cells and clotting proteins have been removed) is added, the rate of cholesterol synthesis is much reduced. Brown and Goldstein demonstrated that the inhibitory factor in human serum is LDL. To inhibit cholesterol synthesis in the cells, the LDL must get in. This work generated the idea of an LDL receptor, a protein that binds LDL and internalizes it in the cells.

Brown and Goldstein repeated this experiment with cells taken from patients with FH. As with cells from healthy people, FH cells show a high rate of cholesterol synthesis in the absence of human serum. However, the rate of cholesterol synthesis is not reduced in FH cells by the addition of serum-containing LDL. The rate of cholesterol synthesis in FH cells is reduced by the addition of pure cholesterol.

These experiments tell us the following. In cells from healthy people, the LDL receptor takes up LDL from the medium, and the associated cholesterol inhibits cholesterol synthesis. The cells from patients with FH must have a defective LDL receptor because LDL itself does not inhibit cholesterol synthesis but the cells still respond to cholesterol itself.

FH comes in two guises. You have two genes, one on each chromosome 19, that encode the LDL receptor, a protein. If one of these genes is flawed and one is normal, you are a heterozygote for FH. This is a rather common genetic disease; about one person in 500 is an FH heterozygote. These people have half the number of LDL receptors and about twice as much LDL cholesterol in their blood as people with two good genes. The consequence is not good; many have heart attacks in their 30s or 40s. If both copies of the gene for the LDL receptor are flawed, you are an FH homozygote, have no functional LDL receptors, and have an extremely high blood LDL concentration. FH homozygotes may have a heart attack by age 10. They are candidates for heart transplants.

There are actually several classes of LDL receptor defects in patients with FH. After all, a mutation could take place anywhere in the gene encoding this protein and there is no reason to think that all the mutations would have the same outcome. Most of the mutant genes do not code for an LDL receptor. There is simply no protein made that has the properties of an LDL receptor. In the absence of LDL receptors, LDL cannot be transported from the blood to the liver. In other cases, a protein recognizable as a mutated LDL receptor is made but cannot find its way to the proper spot on the surface of the vasculature of the liver. The trafficking of the protein within the cell is faulty and LDL is not transported from the blood to the liver. In still other cases, a mutated LDL receptor is made, finds it way to the right place, but does not bind LDL. In all these cases, LDL receptor function is compromised, LDL is not transported from the blood to the liver, and blood LDL levels are too high. Pathology frequently results.

LDL is a particle. Have a look at Figure 8.4. The interior of the particle is largely occupied by a derivative of cholesterol known as a cholesteryl ester. It is made from cholesterol and can be reconverted easily to cholesterol. Surrounding the cholesteryl ester core is a layer of phospholipids and cholesterol. Inserted into the structure is a single, very large protein molecule known as apolipoprotein B-100. The complexes of proteins and lipids such as LDL, known as *lipoproteins*, come in several classes. Some of them are assembled in the liver after a meal. LDL is a metabolic product, formed in the blood, of other, lighter lipoproteins. LDL is the principal, but not the only, carrier of cholesterol in the blood.

The problem is that LDL particles tend to get taken up by cells of the arteries. These lipid deposits build up gradually over time, become calcified, and eventually cross-link with the protein fibrin. The result is a tough plaque buildup in the artery, which is

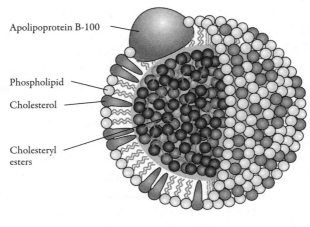

Apolipoprotein B-100

Phospholipid

Cholesterol

Cholesteryl esters

LDL

FIGURE 8.4 Structural Organization of the Low-Density Lipoprotein (LDL) Particle. Note that the particle interior is highly hydrophobic and is surrounded by a somewhat more hydrophilic layer. Just one molecule of the protein apolipoprotein B-100 is complexed with each LDL particle.

known as *atherosclerosis*. The plaque can block partially or completely the flow of blood in the affected artery.

Atherosclerosis can affect a number of arteries: heart, brain, arms, legs, pelvis, and intestines. It can, and frequently does, lead to disease in these organs. For example, if plaque builds up in the arteries of the heart—the coronary arteries—the flow of blood to the heart is reduced, and chest pain can occur (angina) as well as heart attacks.

The carotid arteries provide blood to the brain. Plaque buildup in these arteries may lead to an occlusive stroke. Plaque-dependent restriction of blood flow to arms, legs, or pelvis leads to numbness and pain. Last, atherosclerosis in the arteries providing blood flow to the intestines may lead to pain (abdominal angina) and bowel infarction—a gut attack—which can be fatal. These events are among the leading causes of death in the United States and elsewhere.

There is only one important way to get LDL particles out of the blood and that is via the liver LDL receptor, the protein molecule discovered by Brown and Goldstein. To reduce circulating LDL cholesterol levels, one wants to elevate levels of the liver LDL receptor. Their numbers on liver cells are elevated when cholesterol levels in these cells are lowered, by inhibiting HMGR, for example, which makes sense. When liver cells need more cholesterol for synthesis of bile acids, they generate more receptors to bring in more LDL cholesterol from the bloodstream. Brown and Goldstein capture the whole point in a diagram reproduced here in Figure 8.5.

The LDL receptor is a protein localized to certain regions—termed *coated pits*—on the surface of liver cells (hepatocytes) in contact with the blood. This protein has a high affinity for LDL particles. A blood LDL cholesterol level of only 25 to 30 mg/dL is high enough to saturate the receptor and supply the liver with adequate cholesterol to

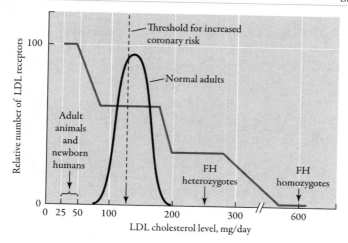

FIGURE 8.5 Blood LDL Cholesterol Levels as a Function of the Number of Liver LDL Receptors. The range of LDL levels in normal adults refers to those in western industrial societies. FH, familial hyperlipidemia. (Source: Redrawn from Figure 16 in Brown, M. S., and J. L. Goldstein. 1985. *A receptor-mediated pathway for cholesterol homeostasis.* Nobel lecture, December 9.)

meet physiological needs. This is approximately the level of LDL cholesterol in the blood of a newborn human baby. Most adults have four to eight times this level of LDL cholesterol, a consequence of a diet rich in cholesterol and saturated fatty acids.

Each LDL receptor can interact with one LDL particle and internalize it. The process is provided in schematic form in Figure 8.6. Once the LDL particle is internalized in a liver cell, the apolipoprotein B-100 protein is broken down to its amino acids and the cholesteryl esters are converted to cholesterol, which may undergo several fates: assembly into lipoproteins, export to peripheral tissue for conversion to steroid hormones, conversion to bile acids, or reconversion to cholesteryl esters that may be stored. The LDL receptor is retained intact and returned to the surface of the liver cell again to interact with an LDL particle.

In addition to supplying the liver with cholesterol to make steroid hormones and bile acids and to participate in membrane formation, two very important things happen when cholesterol levels in the liver are elevated: the level of LDL receptors is reduced and the activity of the enzyme HMGR is reduced. This is a negative feedback regulatory system. In other words, a product of the pathway acts to inhibit the pathway by interfering with an earlier step. This is important; let's think about it for a moment.

Imagine that you are a liver cell. For the reasons that I have mentioned more than once, you need a supply of cholesterol. You have two ways to get your cholesterol fix: you can make it yourself—the enzyme HMGR is the slow step in getting that done—or you can sequester it from the LDL in the bloodstream in which the LDL receptor is the critical molecule.

Now it makes no sense at all to let these processes run wild. As noted previously, it takes 30 chemical steps—and a lot of chemical energy—to make a molecule of cholesterol. So you want to make enough but have no reason to make more than enough. The same

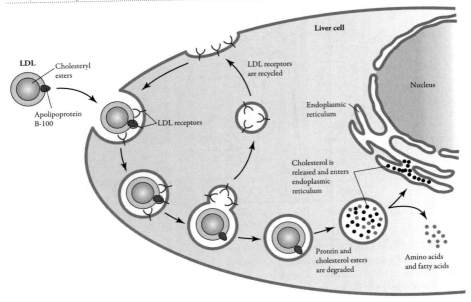

FIGURE 8.6 Sequence of Events in the Low-Density Lipoprotein (LDL) Receptor Pathway in Human Liver Cells. The LDL receptors are localized in the coated pits at the surface of the liver cell. There they may interact with an LDL particle and internalize it. The protein and cholesteryl esters are degraded, but the LDL receptor is recycled and returns to the cell surface.

argument goes for the levels of the LDL receptor. Protein synthesis requires energy and there is no reason to waste it. So as the liver accumulates cholesterol from either source, it makes good sense to reduce the levels of HMGR and the LDL receptor. If the cholesterol level in the liver falls below an appropriate level, then levels of these two proteins will rise again. The key control molecule is an oxidized form of cholesterol itself. By influencing the rate of synthesis of our two key protein molecules, the level of cholesterol in the liver is regulated.

One more point here. As noted earlier, a portion of cholesterol in the liver is converted to bile acids, which are dumped into the intestine to aid in digestion of fats. Although most of these bile acids are reabsorbed in the intestine and returned to the liver, a portion is excreted in the feces. This is the only route of elimination of cholesterol from the human body. This route depends critically on the LDL receptor that brings cholesterol into liver cells, where it is metabolized to bile acids. Statins elevate LDL receptor levels that, in turn, lower blood cholesterol levels. This is all we need to know to understand how the statins work to protect us from atherosclerosis and its consequences.

Elevated Blood Cholesterol Is a Determining Risk Factor for Atherosclerosis

It has been known for some time that high blood cholesterol levels (hypercholesterolemia) and particularly high LDL cholesterol levels are a risk factor for development of atherosclerosis and its sequelae: heart attacks, strokes, and peripheral artery disease. The underlying idea is known as the lipid hypothesis that "postulates that hypercholesterolemia is a major causative factor in atherosclerosis and coronary heart disease."[5] In one form or another, the idea of the lipid hypothesis has been around at least since 1913. It had a stormy history before its general acceptance, with several authorities arguing strongly against it. This fascinating history has been brought together by Daniel Steinberg, one of the key players in the field for decades, in his book *The Cholesterol Wars*.

Too much blood cholesterol has been termed a determining factor for the development of atherosclerosis—that is, it is sufficiently important that reducing blood cholesterol will diminish the burden of atherosclerosis. LDL is recognized as the most atherogenic (causing atherosclerosis) of the blood lipoproteins.

One of the key events in establishing the lipid hypothesis was an interventional trial known as the Coronary Primary Prevention Trial (CPPT) reported in 1984. This study was carried out at several centers of excellence designated as Lipid Research Clinics. This trial enrolled 3,800 men with elevated blood cholesterol levels who were treated for an average of 7.4 years with a cholesterol-lowering agent known as *cholestyramine* (discussed later). In this population, blood cholesterol was lowered by an average of 13.4 percent, LDL was lowered by 20.3 percent, and serious coronary events were decreased by 19 percent. These results, correlating reduced blood cholesterol and LDL levels with reduced coronary disease, were an important factor in organizing the medical community to take action against high blood cholesterol levels.

The Framingham Heart Study, an observational trial, quickly followed the CPPT. In 1948, the National Heart Institute (now the National Heart, Lung, and Blood Institute) of the NIH enrolled more than 5,000 people from the city of Framingham, Massachusetts, in an epidemiological study that continues to this day. The participants return to a study site every 2 years for a detailed examination and history acquisition. In 1971, an additional cohort of more than 5,000 people was added to the study. Over time, a great deal of information has been collected that correlated many factors to the risk of developing cardiovascular disease. Elevated LDL cholesterol is one of them.

The Framingham study links elevated blood LDL levels to the sequelae of atherosclerosis. This association is not the same as proving that lowering elevated blood LDL levels prevents these sequelae, although it surely suggests that this may be the case. As noted earlier, the CPPT provided convincing evidence in support of this hypothesis.

There is a family of independent risk factors for the development of cardiovascular disease. These risk factors include high blood pressure, smoking, diabetes mellitus, lack of physical exercise, obesity, and low levels of blood high-density lipoproteins (HDLs) in addition to elevated blood LDL levels. So LDL is not the whole story, although it is certainly a critical factor. Your blood LDL level is controlled by several factors, but the most important are heredity and diet. There is not much you can do about who your parents are, but you can control what you eat. Diets low in saturated fat and cholesterol contribute to low blood LDL levels. Saturated fat is more effective than cholesterol itself in raising blood cholesterol and LDL levels. However, the main point is that blood cholesterol is the key factor in predicting risk of atherosclerosis and coronary heart disease, not diet itself.

Mevastatin Was the First Statin to Be Discovered

As emphasized back in chapter 5, natural products have been a prominent source of drugs for human health. The statins provide a great example. Back in the 1970s, Akira Endo at Sankyo Company in Japan isolated compactin, since renamed *mevastatin*, from an extract of the fungus *Penicillium citrinum*. The background story is interesting.

Endo joined the Sankyo Company after acquiring his PhD in biochemistry in 1957. His work during the next dozen years yielded several commercial products for Sankyo. In a show of appreciation, Sankyo granted Endo the opportunity to spend 2 years working on a project of his choice. In my experience, this is somewhere between very unusual and unheard of. Endo took full advantage of the opportunity. Allied with his colleague Masao Kuroda, he began a search for inhibitors of cholesterol synthesis. His assay involved the rate of conversion of ^{14}C-labeled acetic acid into cholesterol, following the much earlier work of Bloch, Cornforth, and Popják described earlier. This was not an easy assay, but it provided the opportunity to find inhibitors of any of the 30 enzymes along the pathway from acetic acid to cholesterol.

Endo and Kuroda chose fungal extracts as their source of potential inhibitors of cholesterol synthesis. For nearly 2 years, they screened fruitlessly nearly 6,000 of these extracts. Then, on March 15, 1972, Endo and Kuroda found potent cholesterol synthesis inhibition activity in an extract of *Penicillium citrinum*, a fungus isolated from rice fields. This is an important date in the history of prevention of atherosclerosis and its pathological sequelae.

It took Endo several years to purify the inhibitor, originally known as *compactin* and later named *mevastatin*, as mentioned, to determine its molecular structure and to identify HMGR as the target enzyme. The results of this work were published in 1976.

The central question facing Endo at this point was: would mevastatin prove effective in reducing cholesterol in experimental animals? The answer was by no means obvious. The molecule might not be absorbed from the gut; it might be metabolized quickly to an inactive molecule; it might not localize at the site of cholesterol synthesis, the liver; or a counterregulatory response in living animals might negate the effect of mevastatin. As emphasized back in chapter 5, showing efficacy in experimental animals is a key step on the drug discovery ladder. Endo was able to demonstrate that mevastatin was quite effective in lowering blood cholesterol levels in both dogs and monkeys. In contrast, it was not effective in rats. In rats, most of blood cholesterol is carried in HDL, not LDL, which is present in very low amounts. So even an impressive lowering of LDL would not translate to much of an effect on blood cholesterol.[6] Thus, the critical question for clinical trials was: would mevastatin behave in people as it does in dogs and monkeys or as it does in rats?

Akira Yamamoto, a physician at the National Cardiovascular Center in Osaka, Japan, agreed to collaborate with Endo. He treated 11 patients with high blood cholesterol resistant to the action of cholesterol-lowering drugs then available with mevastatin. The result was spectacular. Mevastatin proved highly effective in reducing the blood cholesterol levels in these patients. This result was published in 1980 in an obscure journal. A second clinical trial confirmed the exciting earlier results and the outcome was published in *Atherosclerosis* in 1980.[7] Another key finding was that mevastatin lowered LDL cholesterol levels without lowering HDL levels, exactly what one would have hoped for. Note that HDL, good cholesterol, is a negative risk factor for the development of atherosclerosis; you want to keep it high, not low. The work of Endo and Yamamoto confirmed the underlying hypothesis of this work—an inhibitor of HMGR would reduce elevated blood levels of LDL effectively in humans. Akiro Endo was awarded the Lasker-DeBakey Clinical Medical Research Award in 2008 for his groundbreaking contribution to human health.

Mevastatin Acts to Raise LDL Receptor Levels in the Liver

How mevastatin works to lower blood LDL levels is not obvious. Brown and Goldstein uncovered the mechanism, based on the regulation of cholesterol metabolism described earlier. Inhibition of the enzyme HMGR will initially inhibit cholesterol biosynthesis in the liver. As noted previously, to overcome this effect, two things happen. First, the cells of the liver make increased amounts of HMGR, which makes sense because increased amounts of this critical enzyme tend to restore the rate of cholesterol synthesis to its normal level. Basically, the liver is fighting back against the enzyme inhibitor. The rate of cholesterol synthesis in the liver is only slightly diminished in the presence of a statin. This is a very good thing because, as detailed earlier, cholesterol is absolutely required for human life. In addition, there are a number of products that branch off from the cholesterol synthesis pathway that are also important (see Figure 8.3). Inhibition of their synthesis might have led to safety issues. Preservation of the rate of cholesterol synthesis also ensures preservation of their rate of synthesis—a good thing.

There is another important point here. Inhibiting HMGR has the potential to cause its substrate, hydroxymethylglutaryl-CoA (HMGCoA), to accumulate in cells, and this could be a safety issue. Fortunately, there is an enzyme that degrades HMGCoA to simpler molecules that are themselves metabolized. This point is not trivial and here is why.

During the mid 1950s, The Wm. S. Merrill Company discovered a compound that was a promising inhibitor of cholesterol biosynthesis. In dogs, it lowered blood cholesterol levels by 20 to 25 percent. The molecule was given the generic name *triparanol* (aka *Mer/29*) and was eventually marketed under that name. Triparanol inhibits the very last step in cholesterol biosynthesis in which desmosterol is converted to cholesterol. These two molecules are related very closely structurally, and this should have raised a red flag. The accumulation of desmosterol posed a threat to human safety because it might substitute for cholesterol. Scientists working with the drug observed formation of cataracts and hair loss in dogs and rats. Some of these scientists, a group from Merck, notified

Merrill of these toxic effects months before triparanol was approved for marketing. Merrill denied having observed toxic effects despite having seen them in their own toxicology studies. They were not reported in the FDA filing in support of triparanol. In 1963, a federal grand jury brought criminal charges against Merrill and some of its employees. Merrill eventually paid a fine of $50 million. This unfortunate episode set back the search for effective and safe inhibitors of cholesterol biosynthesis as others worried about consequent toxicities. Which step in the biosynthetic sequence that you choose to inhibit matters.

As noted earlier, inhibitors of HMGR cause the liver to synthesize more of this enzyme. The second thing that happens is that the liver cells increase the synthesis of LDL receptors, which makes sense because more LDL receptors permit the liver to sequester more cholesterol from the blood. This, too, is a very good thing, because getting LDL out of the blood is the whole point. An increased number of LDL receptors lowers the blood LDL cholesterol level under the action of statins (see Figure 8.5).

This is a beautiful result. Think back to the case of FH heterozygotes. Their problem is a genetic defect that lowers the number of LDL receptors. A statin tends to restore the number of LDL receptors to normal levels. It does the same for others with high LDL cholesterol levels. Thus, the mechanism of action of the statins augments a natural control system.

Merck Enters the Race

Based on the results at Sankyo Drug Company with mevastatin, one thing was clear: the statins had the potential to be blockbuster drugs. There were drugs on the market that lowered blood cholesterol levels. Perhaps the most interesting was cholestyramine, marketed as Questran. This is a bile acid sequestrant. Cholestyramine binds to the bile acids in the gut and prevents their reabsorption. Remember that bile acid formation is a major route of metabolism of cholesterol, but most of the bile acids dumped into the gut are reabsorbed. By preventing this, cholestyramine accelerates the rate of clearance of cholesterol from the body, which induces the liver to ramp up its synthesis of the LDL receptor to compensate. The effects are summarized in Figure 8.7. Note that the final mechanism of action of statins and cholestyramine is the same: increase the LDL receptor level.

There are two problems with cholestyramine. First, it does not work very well. Most patients on this drug achieve a 15 to 20 percent lowering of LDL cholesterol. This is a positive result, but it is not a large enough effect to get the cholesterol level where it should be. Second, cholestyramine is a gritty resin that must be taken in large quantities, perhaps 20 g a day. Many patients have an adherence problem with this regimen. This issue has been overcome, to some extent, by the formulation of cholestyramine and other bile acid sequestrants in tablet form. There were other options available, but

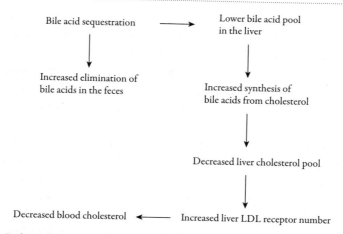

FIGURE 8.7 Pathway from Sequestration of the Bile Acids by Cholestyramine or Similar Agent to Reduction of Blood Cholesterol Levels. Note that both bile acid sequestrants and statins have the same final outcome: an increase in the number of liver LDL receptors.

none very satisfactory. The statins promised better things. Finding one quickly became a high-priority research objective at Merck (and surely at a number of other pharma houses as well).

This effort began at Merck before I joined the company.[8] Merck chemists had begun work to synthesize HMGR inhibitors in late 1977. This work did not prove promising, and attention was subsequently directed to a search for these inhibitors in natural products, as described next.

From my years in academia, I knew the cholesterol biosynthesis story, but not a whole lot more relevant to statin discovery. So a lot of learning was in store for me. Roy Vagelos had cleverly reached out to Mike Brown and Joe Goldstein as consultants on the project. So, we had the two world authorities on LDL receptors and cholesterol metabolism to whom to turn for advice. Both became intensely committed to bringing a statin into clinical use. I had two connections with the Merck statin effort.

First, I was asked to be the Discovery Project Committee chairman for the statin effort. When a discovery project gets well under way and shows some signs of promise, a committee is assembled to coordinate the effort. Each committee has a representative from each function required to progress the work—typically, a chemist, a molecular biologist or biochemist, a drug metabolism expert, and a pharmacologist to start. Others are added as the project moves forward, perhaps a toxicologist and a clinician. The committee also has the services of someone from Project Management, a unit with the task of seeing that all the threads required to move things forward without loss of time are in place.

Second, Alfred W. Alberts led the group that did all the basic biology associated with the statin discovery effort. Al reported to me. Al had trained with Roy Vagelos at the NIH, moved with him to Washington University in St. Louis, Missouri, and moved with Roy again when he joined Merck. He was fully capable of doing what needed to be done on the biology side to move things forward. He was particularly good at devising clever experiments to shed light on issues as they arose. He developed the critical assay that measured the potency of HMGR inhibitors, a critical property, because increased potency reduces dose, which reduces risk along with drug cost. This assay was actually run by his talented technician Julie Chen. In a very real sense, Al led the effort to discover statins for Merck.

On the chemistry side was Art Patchett, the same guy who discovered the two Merck ACE inhibitors. Art was director of the New Lead Discovery department that synthesized mixtures of compounds for testing in novel biological assays under development at Merck at the time. In 1974, he also set up a procedure in which fungal and other microbial fermentation broths were extracted, concentrated, and sent for testing in biological assays. The procedure was low capacity, so cultures were chosen based on interesting morphology or if there were other reasons to suggest notable biological activity. No major discoveries were made from this screening system until fall 1978. Earlier, Al Alberts had established his assay for inhibiting HMGR. He and Julie Chen began to screen Art's extracts.

In an extraordinary stroke of good fortune, the 18th broth extract proved to be superbly active. In fact, it contained what we now call *lovastatin*, a direct product of the metabolism of *Aspergillus terreus*. In contrast, Endo's group had examined 6,000 cultures before they discovered and isolated mevastatin. Months would have elapsed had Merck waited to do high-volume HMGR screening in their Spanish fermentation screening laboratories. Instead, the New Lead Discovery department's broth extract yielded readily pure lovastatin.

Carl Hoffman's group in the New Lead Discovery department had prepared and sent the active broth to Al Alberts. Carl's group also succeeded in isolating pure lovastatin from the fermentation extract.[9] In a tour de force of wit and effort, Georg Albers-Schönberg at Merck determined the structure of lovastatin over a weekend. He received the isolated molecule on a Friday and reported the structure at a meeting the following Monday. It proved to be very closely related to that of mevastatin (Figure 8.8).

Both structures resemble, in part, that of the natural substrate of HMGR, providing a sensible basis for the inhibitory activity. Lovastatin and other statins fit into the active site of HMGR, thus preventing access by the natural substrate (Figures 8.9 and 8.10). Al Alberts and his group established that it was a highly potent inhibitor of HMGR and lowered LDL cholesterol levels dramatically in dogs. Merck turned on a major drug discovery effort to understand lovastatin in as much detail as reasonably possible. Merck's efforts in support of lovastatin are described by Roy Vagelos as follows:

> In addition to chemists and microbiologists, the growing team now included chemical engineers, spectroscopists, pharmacologists, and toxicologists. The microbiologists determined the optimal conditions for growing the microorganism. The chemical

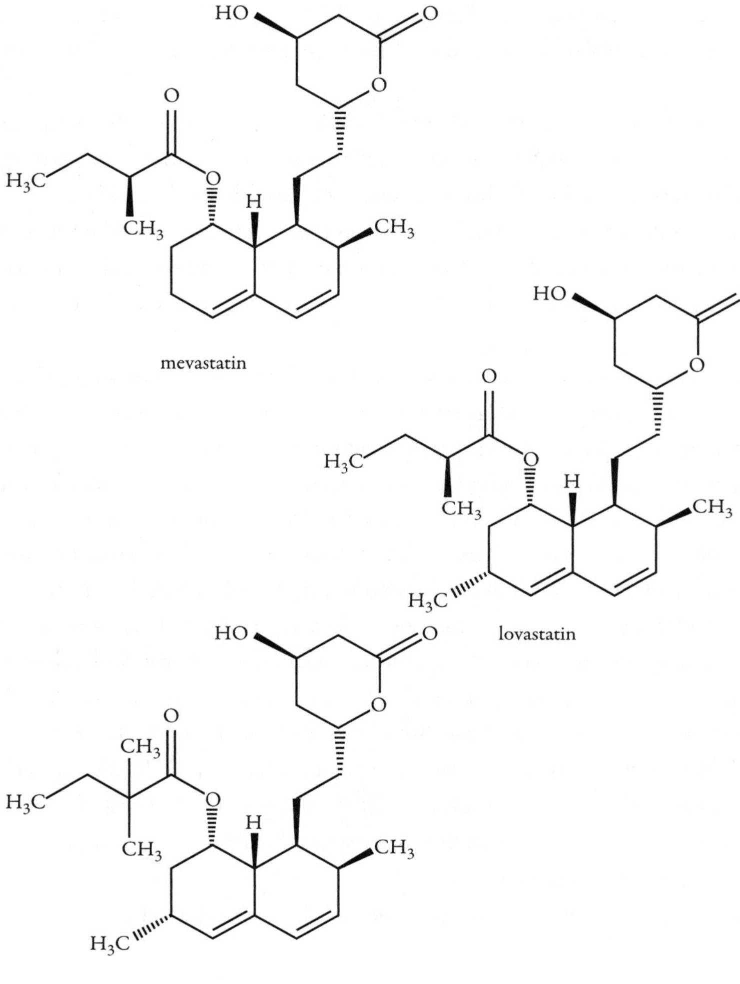

mevastatin

lovastatin

simvastatin

FIGURE 8.8 Structures for Mevastatin, Lovastatin, and Simvastatin.

engineers isolated the lovastatin in large quantities. Spectroscopists determined its chemical structure. Pharmacologists studied its effects in live animals, and the toxicologists studied lovastatin to demonstrate any possible harmful effects by feeding it to mice, rats, and rabbits. We also started to assemble a clinical research team and altered marketing.[10]

Lovastatin proved to be a marvelous molecule. It is a potent and specific HMGR inhibitor and is profoundly active in lowering blood cholesterol in dogs and monkeys. It is also well absorbed when given to animals orally and it has a long duration of action, as judged by its ability to prolong the cholesterol-lowering action. Furthermore, the great majority of lovastatin given orally localizes in the liver, the site of cholesterol synthesis. The lovastatin project quickly became the highest priority discovery effort in Merck Research.

In anticipation of clinical trials, short-term safety assessment studies with lovastatin were begun under the leadership of Jim MacDonald. Lovastatin proved to have a good safety profile in these studies, fully consistent with short-term clinical trials.

An IND was filed with the FDA and clinical trials were initiated. As expected, lovastatin proved to lower blood LDL levels dramatically in people and revealed a good safety profile. The development of lovastatin was moving nicely and Merck scientists were excited.

In addition to moving lovastatin along in clinical trials as rapidly as possible, Merck started a medicinal chemistry effort to modify lovastatin in an effort to find a backup or second-generation molecule. This is pretty much standard practice in the pharma industry. If something should go wrong with development of the first molecule, and it frequently does, it is best to have a second or third molecule to bring along behind it so you are not out of the game entirely. This work was done under the leadership of Paul Anderson and Bob Smith at the West Point, Pennsylvania, site of Merck Research.

Last, in an effort to save time in the long run, carcinogenicity trials were also started. These are the longest safety assessment studies required for approval of a drug for marketing and they determine the potential of a compound to produce cancer. Rats and mice of both sexes are treated with the highest dose of the test drug they can tolerate for their lifetime (about 21 months in mice and 24 months in rats). It takes an additional 6 months, more or less, to examine every body tissue for signs of cancer. So you are looking at 2.5 years from the start of carcinogenicity trials until they are complete, and additional time to generate a final report.

Everything was moving smoothly, then we hit a bump in the road.

Mevastatin Fails

I have yet to see a drug discovery and development program that did not appear dead at some point in its evolution. The lovastatin project came to a complete halt and it was unclear whether it could be rescued. Here is why.

Sankyo Drug Company suspended the development of mevastatin, reportedly because the drug caused intestinal tumors in dogs. This caused great concern at Merck because (1) mevastatin and lovastatin work by the same mechanism and (2) the structures of mevastatin and lovastatin are very closely related. So whether the induction of tumors was a consequence of mechanism or structure, lovastatin appeared to be in big trouble. If mevastatin caused tumors, it seemed likely that lovastatin would cause them, too.

The Sankyo announcement happened to come while the leadership of Merck Research was gathered for a four-day annual meeting in southern New Jersey. The meeting reviewed progress on projects during the past year, a preliminary to making resource allocations—people and dollars—for the upcoming year. It was taken seriously. The response of Roy Vagelos was immediate; he halted the clinical trials of lovastatin based on a legitimate fear for the safety of trial participants. Beyond that, he invited to join us those scientists in the research organization who might be able to help us think through the problem. We met and discussed various issues for several hours until late in the evening. Among other matters of importance to come out of the meeting was a decision to start a new medicinal chemistry program at the West Point site of Merck Research that was designed to discover an inhibitor of HMGR that was distinct structurally from mevastatin and lovastatin. Bob Smith was chosen to lead the effort. Bob was a highly capable medicinal chemist and an irrepressible optimist. He assembled a large group of chemists and work began.

Sankyo was not forthcoming with critical information relevant to the rumored carcinogenicity of mevastatin. We would like to have known which species, which kind of tumors, which dose, and following which period of treatment. In the best case, this information might have permitted Merck to do a short-term study to determine whether lovastatin behaved similarly, but we did not get this information. Rumors suggested that the issue was gastrointestinal tumors in dogs after 6 months of treatment, but rumors are just that—and we did not find out. Roy Vagelos spoke directly with his counterpart at Sankyo without result. Roy also traveled to Japan to meet with Sankyo leaders on the issue, also without result.

Shortly thereafter, Merck Research management made an error in judgment. Contrary to the wishes of the Project Team, management decided to terminate the ongoing carcinogenicity studies with lovastatin. This was an understandable decision but a flawed one. We were about a year into the studies and having to start anew would cost at least 2 years and would prevent us from generating information to enable an informed decision on the safety of this whole class of drugs. Happily, the leader of Safety Assessment at Merck at the time, Del Bokelman, with the urging of Jim MacDonald, who was responsible for the lovastatin safety assessment work, elected to ignore the management decision and completed the studies.

Although these studies yielded an increase in spontaneous tumors in several tissues, for a variety of sound scientific reasons this response was judged not to represent a risk for humans at therapeutic levels. Global authorities agreed with this judgment by Merck scientists. Completing the carcinogenicity studies saved Merck a couple of years in getting to the marketplace with lovastatin. I do not recall that either Bokelman or MacDonald ever got credit for ignoring the wishes of management, but they should have.

Bob Smith and his people were successful in finding HMGR inhibitors that were distinct structurally from lovastatin. However, none of them fulfilled the family of criteria required for a move into development: potent cholesterol-lowering activity, good oral activity, and long duration of activity in experimental animals. In short, they were not as good as lovastatin.

Three prominent cardiologists—Roger Illingworth, Scott Grundy, and David Bilheimer—all Merck consultants and all frustrated by the lack of a potent agent to treat high LDL blood levels in their patients argued for a limited restart of lovastatin clinical trials, which was strongly supported by Mike Brown and Joe Goldstein. The decision to do so was not easy. Merck carcinogenicity trials were not complete, and no useful information about the safety issues with mevastatin was forthcoming. Subsequently, Daniel Steinberg and Jean Wilson, both authorities in the field of cholesterol regulation, and both members of Merck's Scientific Advisory Board, also supported a restart of clinical trials. Roy Vagelos agreed. Lovastatin was back on track.

Although the effort of chemists at the West Point site to find attractive HMGR inhibitors that were distinct structurally from lovastatin did not succeed, they did come up with a big winner: the discovery of a modest modification in the structure of lovastatin that created a more potent inhibitor of HMGR. It proved better than lovastatin at lowering blood cholesterol in animals without sacrificing any of the important properties of that molecule. It was named simvastatin (see Figure 8.8), later marketed as Zocor, and was the biggest product in the history of Merck.

Safety Issues Loom for Lovastatin and Simvastatin

Long-term safety assessment studies of lovastatin in experimental animals turned up a family of six issues at high doses of the drug that required resolution, in addition to the carcinogenicity findings mentioned earlier. Each of these was approached in a data-driven experimental manner in an effort to judge their potential impact on human safety. By way of illustration of what measures were taken, let's consider two of them— liver toxicity in rabbits and cataracts in dogs—one at a time.

The cataract issue was observed initially in one dog only on the final day of a 1-year safety assessment test and only at the highest dose tested, 180 mg/kg (the effective human dose is less than 1 mg/kg). At this dose of lovastatin, blood cholesterol in dogs is lowered dramatically. However, this singular finding raised a red flag and the study was continued. Additional examples of cataracts in dogs at very high doses of lovastatin were observed.

The observation of cataracts in dogs was handled by conducting eye examinations (specifically, slit-lamp examinations in search of evidence of cataracts) in all people enrolled in clinical trials. The result was clear: at effective doses of lovastatin taken long term, there was no evidence of cataract formation. This safety issue was laid to rest.

There are two ways to think about the liver toxicity observed in rabbits. First, this finding might have been mechanism based. That is, it might result from the action of the drug. In this case, all statins would have the same toxicity liability. At the high doses used in the safety assessment studies, blood levels of cholesterol decreased dramatically (rabbits have a very low HMGR content). As established at the beginning of this chapter, cholesterol is absolutely required for good health (and life) and it would not be surprising if signs of toxicity showed up when the levels of this substance are very low.

Second, there were several possibilities other than those that are mechanism based, including action of lovastatin and simvastatin at targets other than HMGR, formation of one or more toxic metabolites, or something that we had not yet thought of. Mechanism-based toxicity can sometimes be approached rationally. Everything else poses a really difficult problem. Here is how to think about the mechanism-based toxicity for a statin.

The mechanism of action of the statins is to inhibit HMGR. The action of this enzyme produces mevalonate. At very high doses of a statin, one expects that levels of mevalonate, and other metabolites on the route to cholesterol, would be depressed. There is a straightforward way to know whether this underlies the toxicities: feed mevalonate to experimental animals treated with statins. This strategy would restore mevalonate levels and those of metabolites on the route to cholesterol. The idea is captured in Figure 8.11.

If the liver toxicity observed in rabbits was mechanism-based, then mevalonate replacement should eliminate the toxicity. Basically, feeding a statin and mevalonate together is a lot like doing nothing, because the mevalonate neutralizes the effects of the statin. The result of the mevalonate replacement study was clear: mevalonate prevented hepatotoxicity in rabbits. This established that the toxicity was mechanism based and reflected the very low levels of blood cholesterol that result from very high doses of the statin. Because the point of treating patients with a statin is to get their elevated cholesterol back to within a *normal* range with a modest dose of the drug, this safety hurdle was laid to rest as well.

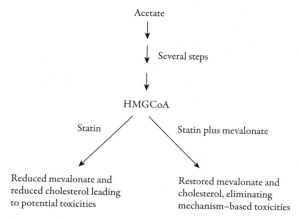

FIGURE 8.11 Mevalonate Replacement Experiment. Here is the rationale for mevalonate replacement experiments as a test for mechanism-based toxicity of statins. Mechanism-based reductions in mevalonate, cholesterol, or intermediates may, if sufficient, create toxicity. The addition of mevalonate to the statin neutralizes the effect of the statin, normalizes mevalonate and cholesterol levels, and prevents toxicity.

Lovastatin and Simvastatin Succeed

Under the leadership of Jonathan Tobert, the clinical program for lovastatin went smoothly. The molecule proved highly effective in lowering LDL cholesterol at a modest oral dose taken once a day. Simvastatin followed.

Along about this time, Roy Vagelos succeeded John Horan as CEO of Merck. These were great days for Merck. The research laboratory moved several important drugs into clinical use and profits soared, as did the stock price. Merck was named The Most Admired Company in the USA for 7 consecutive years. Roy's work as president of Merck Research was a very important factor in that success, and his achievements were widely admired. On Roy's eventual retirement from Merck, the Rahway, New Jersey, site of Merck and Company was renamed the P. Roy Vagelos Research Center. Incidentally, Roy was a graduate of Rahway High School.

Ed Scolnick, who had been recruited to Merck Research from the NIH by Vagelos, was named the new president of Merck Research. Ed is a molecular biologist and an outstanding scientist. He is not only a very bright guy, but is perhaps the most focused and driven person I have met. Somewhere along the way, I had been promoted to vice-president of Biochemistry and Molecular Biology, Ralph Hirschmann had been reassigned, and I now reported to Ed. We had a good working relationship and I learned a good deal from him, as I had from Ralph, and my job was always interesting.

Ed took a great interest in the lovastatin/simvastatin work and became, in reality, the leader of the effort. My role had, at any event, diminished greatly when lovastatin entered clinical trials. My task was on the discovery side, not the development side. As the program moved forward, Eve Slater took an increasingly important role, particularly in helping resolve the safety issues described earlier.

In early 1987, an NDA was assembled for lovastatin and submitted to the FDA. Subsequently, Merck made a presentation to the Cardiovascular Advisory Committee of the FDA. Eve Slater made the presentation along with Mike Brown, who was there in support of lovastatin. I missed the presentation; it was my last day at Merck. I had made a decision to become president of research and development for another pharma company, which worked out well, but is another story.

Merck's presentation was very well received. The Advisory Committee was unanimous in its recommendation for approval. Six months later, in 1987 (a remarkably short time), the FDA granted Merck the right to market lovastatin as a cholesterol-lowering agent under the trade name Mevacor. In 1988, simvastatin was approved for the same indication and was marketed under the trade name Zocor. Both proved to be blockbuster drugs, with annual sales in excess of $1 billion.

There remain a few key points. The activity of lovastatin and simvastatin to lower cholesterol levels was the basis for their approval for marketing. At the highest doses used in clinical studies, lovastatin reduced blood cholesterol by 40 percent and simvastatin lowered it by 47 percent. However, they also lower blood levels of triglycerides and very low-density lipoproteins, the principal carriers of triglycerides in the blood. Triglyceride levels alone are not a significant risk factor for atherosclerosis. However, in the context of elevated blood cholesterol, elevated triglycerides do become a risk factor and lowering them is good. These statins have a modest effect of raising levels of HDL, good cholesterol—another positive point. Lowered blood cholesterol levels were taken as a surrogate for protection from atherosclerosis. So at the time of their entry into clinical practice, there was no evidence of efficacy based on the clinical end points of heart attacks, strokes, and peripheral artery disease. These necessary data were acquired after approval in three waves. First, statins were shown to be effective in secondary prevention trials—that is, for patients who had already suffered a heart attack or stroke. Second, statins were shown to be effective in primary prevention trials—that is, for patients who had not yet had a heart attack or stroke but had elevated cholesterol levels and were at risk. Last, statins were shown to be protective for patients with cholesterol levels that fall in the range considered normal.

Biomedical researchers have now had the statins for 25 years, and a great deal of clinical information has been obtained. All told, statins have been evaluated in at least 14 major clinical trials enrolling close to 100,000 patients. In a 2012 editorial in the *New England Journal of Medicine*, Allison Goldfine summed up the results:

> There is no doubt that for persons who have had an acute coronary syndrome or who have other risk factors for atherosclerotic coronary artery disease, statins effectively reduce the risks of death from any cause, deaths due to cardiovascular disease, fatal myocardial infarction, the need for revascularization, and stroke. Over a period of 4 years of statin use, a reduction of 1 mmol (39 mg/dL) in the level of low-density lipoprotein (LDL) cholesterol translates into a 9 percent reduction in the risk of death from any cause among patients with diabetes and a 13 percent reduction among those without diabetes. . . . Few drugs have had such a dramatic effect on health outcomes.[11]

Nothing is perfect. The statins as a class have a rare but potentially serious adverse effect on muscles termed *rhabdomyolysis*. Like all toxicities, this one is dose and exposure related. It is generally true that a dose can be found that is free of a toxic response and permits safe therapy. In addition, retrospective studies have established that use of statins is associated with an increased risk of contracting adult-onset diabetes, although it is not clear why. Rare or subtle adverse effects may be found in long-term clinical safety studies in patients after marketing approval or may come to light only after tens or hundreds of thousands of patients have been treated by their physicians. It takes years of clinical experience with a new drug before the full profile of efficacy and adverse effects are understood. On the positive side, statins have proved to be anti-inflammatory—not anticipated or completely understood—but welcome nonetheless, and realized after years of use. In fact, some authorities have argued that a significant fraction of the therapeutic benefits of the statins are a consequence of their anti-inflammatory properties.

For myself, I am happy to have had a small role in helping lovastatin and simvastatin along the way to clinical use. I am reminded of a story told by the great Hungarian biochemist Albert Szent-Györgyi after a flowery introduction to one of his talks. It goes like this. A small mouse went down to the ocean and passed his urine into it. Having finished, he turned around and had a look. "Every little bit helps" he said.

9 The Perils of Primaxin

THE *PERILS OF Pauline* is a 1914 film serial in 20 episodes. In each episode, a villain, perhaps a pirate, menaces Pauline, played by Pearl White. In each episode, Pauline seems certain to meet her demise, only to escape or be rescued at the last possible moment. Outdoing the proverbial cat of nine lives, Pauline had 20.

The *Perils of Pauline* bears more than a passing resemblance to drug discovery and development in which some villain, perhaps an issue of safety or efficacy, threatens the life of a project. I know of no better example of this than the course of getting the antibiotic Primaxin from the laboratory to the bedside. The perils of Primaxin plays out in a scientific serial of five episodes in which the project is rescued from impending disaster in each episode.

The Primaxin story is one of the great tales of drug discovery in the world of antibiotics—the molecules that have power to prevent or cure bacterial infections. Primaxin is a triumph of the pharmaceutical industry in general and of Merck specifically. Victory did not come easily. The road from a drug discovery start to a marketable human health product is often rough, occasionally very rough, and sometimes impassable. The Primaxin story stands out for the number and nature of bumps and ruts that impeded passage. That Primaxin made it to the market for the benefit of countless patients with infectious diseases is a tribute to the wit and determination of the many scientists who saw it through. That story comes later. Let's get started with some general background on the field of antibiotics.[1]

Cemeteries Can Be Useful Sources of Insight and Information

The next time you have the occasion to explore a cemetery, have a look at the tombstones for people born around 1900. A significant fraction of those tombstones will reveal that the date of death is within 10 years of the date of birth. Simply put, many people born around 1900 or earlier did not live to be 10 years old. Now have a look at the tombstones for people born in 1940 or later. A much smaller fraction of people born in that era died before the age of 10. One goal of this chapter is to explore and explain the reasons for this striking difference, with emphasis on the discovery and development of antibiotics.

Great Progress Has Been Made Against Infectious Diseases

Life expectancy in the United States increased markedly during the 20th century, from about 46 years in 1900 to about 76 years in 2000. There are many reasons for the addition of three decades to the average duration of life. Among them are better public health measures, healthier lifestyles (better diets, more exercise, avoidance of tobacco, moderation in the intake of alcohol), availability and use of vaccines, improved surgical procedures, novel technologies for diagnosis of disease (such as magnetic resonance imaging, ultrasound imaging, computed tomography), and powerful new drugs, specifically including several classes of antibiotics. These and other innovations have not only increased the duration of life markedly, but also have contributed to a better quality of life. Here is one way to think about what we are trying to achieve: to remain healthy until the day you die and to postpone that day as long as reasonably possible.

Not only has life expectancy in the United States and elsewhere in the developed world increased dramatically during the past century, but also the nature of life-terminating events has changed as well. In 1900, major causes of death in the United States included infectious diseases such as smallpox, diphtheria, scarlet fever, pneumonia, tuberculosis, and others. These are the diseases that, by and large, are responsible for the frequency of early deaths in people born around 1900, as mentioned earlier. For the most part, people born in 1900 in the United States did not live long enough to die of heart disease or cancer, which are now our major killers. We have made notable progress on the battle against infectious diseases. The continuing fight has found new targets. Many of the major infectious diseases of 1900 are not those of the 21st century. Let's have a short look at key historical events leading up to the Primaxin story.

The European bubonic plague of 1347 killed one-third of the population of Europe. It is the largest single plague ever recorded. The disappearance of the Aztec civilization in the New World was spurred by smallpox and measles introduced by Hernando Cortés and his band of Spanish invaders. The same diseases also decimated Native Americans in what is now the United States. More recently, the influenza epidemic of 1918 killed an estimated 40 million people worldwide. Malaria continues to be a major problem.

AIDS, tuberculosis, influenza, hepatitis, pneumonia, and a lengthy list of parasitic infections continue as important constraints on the welfare of people throughout the world.

Infectious diseases can be fought effectively. Smallpox has been wiped out on Earth; the last case was recorded in Ethiopia in 1977. Rinderpest, a viral disease that affects cattle and other ruminants, has been eradicated. The incidence of poliomyelitis on Earth has been reduced by 99.9 percent during the past few decades, and there are reasonable, short-term prospects for its total elimination. As we learn in chapter 10, there are good prospects for the elimination of the parasitic diseases onchocerciasis (river blindness) and filariasis (elephantiasis). Many of the important infectious diseases of the early 20th century in the United States have been brought under effective control—smallpox, scarlet fever, whooping cough, and others. The discovery and use of novel vaccines and antibiotics during the past few decades have been enormous boons in the fight against infectious diseases. We now live longer and live better because of them.

However, the war against infectious disease continues and will continue for the foreseeable future. Here are three reasons why. First, drug-resistant organisms continue to develop. Methicillin-resistant *Staphylococcus aureus*, MRSA, is one important example. MRSA is resistant to a host of potent antibiotics. The drug of last resort against MRSA and other life-threatening microbes has been vancomycin. Vancomycin-resistant MRSA has been discovered during the past few years, so the last barrier has been breached. It seems reasonable to conclude that resistance will develop to any antibiotic that we come up with in the future. A final solution to the infectious disease problem is going to require something quite new indeed—a technology well beyond what we can envision today.

Second, we have the issue of bioterrorism. There are many potential threats, including smallpox, anthrax, tularemia, plague, and a family of viral hemorrhagic fevers, among others. In addition, there is a limitless number of virulent bacteria that might be constructed using the modern techniques of genetic engineering.

Third and last, new infectious organisms arise continually. We do not understand the microbial diversity on Earth, and it expands.

Truly novel antibiotics are tough to find. Only two antibiotics based on a novel chemical skeleton have been placed into clinical practice during the past 20 years. The discovery of novel antibiotics effective against resistant microorganisms is not a trivial task. Let me provide one example why. A group of German scientists evaluated 700 enzymes, each a potential target for antibiotic development, in the pathogen *Salmonella enterica*. More than 400 of these enzymes proved to be nonessential for virulence in this organism, narrowing the number of potential targets. Of the fewer than 300 enzyme targets that are essential for *Salmonella* virulence, 64 were identified as being conserved in other important human pathogens. These 64 enzymes are, therefore, potential targets for drugs that would have a useful spectrum of activity across pathogenic bacteria. Discouragingly, almost all of them were found to belong to metabolic pathways that are targeted by antibiotics now in routine clinical use. The scope of the field of activity in the search for novel antibiotics has been narrowed substantially by earlier discoveries.

There Are Historical Antecedents to Modern Antibiotics

Of critical importance was the work of Louis Pasteur in France, Robert Koch in Germany, and Joseph Lister in England during the 19th century that established the germ theory of disease. That critical discovery took an amazing amount of time to catch on. Too many people, including scientists, do not understand that some fraction of what they know is just not so. People hold on to their misinformation with remarkable tenacity, retarding the acceptance of new knowledge.

NATURAL PRODUCTS PROVIDED SOME OF THE EARLIEST EFFECTIVE DRUGS

The little black bag of physicians did not have a whole lot of useful medicines in 1900. The role of the physician at that time was diagnosis and prognosis far more than therapy. Quinine, morphine, salicylic acid, digitalis, and ephedrine were known and used in 1900, although their liabilities in the treatment of human disease were not fully appreciated. These medicines are all botanical products with benefits for human health that were known before the advent of medicine and chemistry as we know them. Natural products made an early and favorable impact on human health.

At the same time, vendors of proprietary medicines took advantage of our love of the natural to peddle a variety of ineffective plant-derived products. These products were advertised as safer and more pleasant than those used typically by physicians. For example, Mrs. Winslow's Soothing Syrup and Kopp's Baby Friend had morphine sulfate as their basic ingredient. Hostetter's Bitters was a 78-proof (39 percent ethanol) cocktail. This theme is being replayed currently as alternative medicine. This time around, controlled clinical trials are being carried out, slowly, to establish what works and what does not, and to determine the associated safety profiles. For example, a placebo-controlled, double-blind clinical trial has established beyond reasonable doubt that *Echinacea* is of very limited or no utility for the prevention or treatment of the common cold. Likewise, saw palmetto has been found to be ineffective for prostatism. This is not to argue that alternative medicines have no role in human health. It does, however, establish the need for careful evaluation of such products that, through political interference, have been placed outside the purview of the FDA in the United States.

During the 19th century, chemists had a good deal of success in isolating and purifying natural products from plant sources. Morphine was isolated as a pure compound from crude opium in 1804. Quinine, introduced from the New World as a treatment for fever, was isolated in pure form by 1820 from the bark of the cinchona tree. However, its effectiveness against malaria was soon discovered and it found an alternative, highly important medical use. Sodium salicylate was isolated from the bark of the willow tree in 1821 and was also shown to have analgesic, antipyretic (fever-reducing), and anti-inflammatory properties. It took an additional 76 years, until 1897, to synthesize the acetyl derivative, acetylsalicylic acid, commonly known as *aspirin*—one of the great drugs of all time.

These and other discoveries of biologically active substances in plant material are one source of modern medicinal agents. Natural products continue to provide us with important medical advances. Important examples in the field of antibiotics follow.

THE DYE INDUSTRY OF EUROPE PROVIDED ANOTHER SOURCE OF DRUGS

The other source of modern drugs was the European dyestuff industry of the 19th and early 20th centuries. The goal of this industry was to make useful dyes, principally for fabrics. During the course of handling novel molecules, scientists occasionally make unexpected observations (serendipity) that suggest novel uses. For example, the commonly used sweetener aspartame was discovered by accident when a scientist licked a finger containing a bit of this substance. It turned out to be surprisingly sweet.

Paul Ehrlich, a physician and microscopist who experimented with dyes to stain and thus make visible microorganisms, was the first to test systematically the treatment of syphilis by dyes found to stain the causative *Treponema*. From this work came Salvarsan, and also the founding concept of drug action: to be active a drug must bind to its target (*Corpora non agunt nisi fixata*). Ehrlich is justly regarded as the father of the modern pharmaceutical industry. Penicillins are now the drugs of choice for syphilis. Chemists discovered two other medicinal agents during the early years of the 20th century: tryparsamide for trypanosomiasis, a parasitic disease, and oxophenarsine, also for syphilis.

Subsequently, other drugs have been developed for the same uses that provide advances in efficacy and safety. Nonetheless, these molecules were important at the time of their discovery and were key steps down the path to improved agents. Chemistry has been the other great source of antibiotics.

Sulfa Drugs Were the First Important Antibiotics

The first really useful antibiotics were the sulfonamides or sulfa drugs. These agents were discovered during the 1930s as an outgrowth of a search for therapeutically useful dyes. Specifically, Gerhard Domagk found that a red dye known as *prontosil rubrum* was an active antibacterial agent in mice and rabbits, protecting both species from otherwise lethal doses of staphylococci and hemolytic streptococci. For this key observation, Domagk was awarded the Nobel Prize in Physiology or Medicine for 1939.

Domagk did not stop with his findings in mice and rabbits. The big question was whether prontosil rubrum would be an effective antibiotic in human beings. Unexpectedly, his daughter contracted a severe streptococcal infection. As a desperate measure, Domagk gave her a dose of prontosil rubrum and she made a complete recovery. Domagk did not report this finding until some years later, when others confirmed efficacy in clinical studies.

Domagk also observed that prontosil rubrum was ineffective against bacteria in test tube studies, which was an unusual finding. In current practice, activity at the molecular or cell level is generally a required prerequisite to a search for activity in an animal model of disease. Many current drug discovery protocols would not have found the antibacterial activity of prontosil rubrum. Research revealed that prontosil rubrum is metabolized in the mouse by liver enzymes to a simpler molecule—sulfanilamide.

Sulfanilamide proved to be a safe and efficacious antibacterial agent with antibacterial activity that has been confirmed in the test tube as well as in animals and humans. Prontosil rubrum is a prodrug—that is, a molecule itself devoid of the desired biological activity but which is metabolized to one or more compounds in a living organism that do have this activity. The age of wonder drugs had dawned. Gerhard Domagk was the father of the sulfa drug revolution. A huge family of derivatives of sulfanilamide was subsequently synthesized, and several of them came into medical practice.

Structurally, sulfanilamide is the simplest of the sulfa drugs. It is the parent compound of the family of sulfa drugs. All the others were derived from sulfanilamide by replacing one hydrogen atom of sulfanilamide by a variety of suitably decorated ring structures (Figure 9.1).

Although modern science and experience with what works and what does not has given those involved in drug discovery a leg up, the process of searching around for new molecules by replacing one group of atoms for another continues. This is not a random process. A wealth of experience gained throughout the years has encouraged medicinal chemists to favor some groups over others for reasons of efficacy, safety, stability, duration of action, or other key properties. The availability of high-resolution structures, largely from X-ray diffraction studies, of the molecular targets of drug

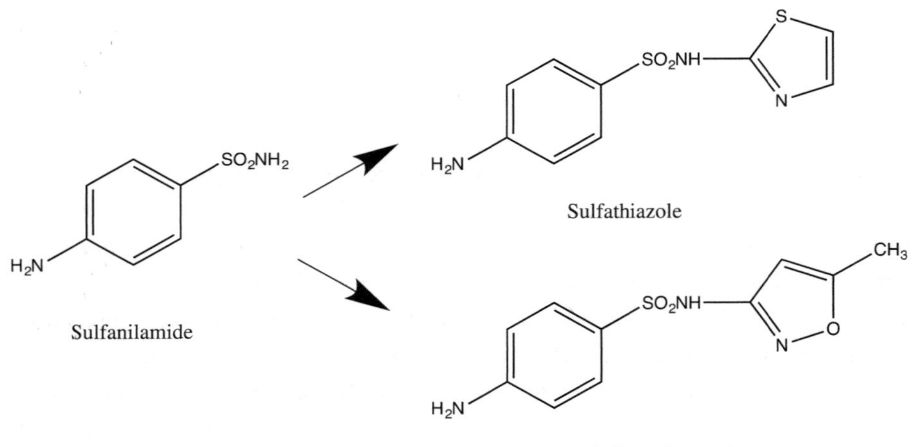

FIGURE 9.1 Sulfanilamide Is the Parent Molecule of the Sulfonamide Class. Sulfanilamide is the first antibiotic discovered in the sulfonamide class. It is also the parent molecule of the large family of sulfonamide antibiotics discovered subsequently. Two examples are provided here—sulfathiazole and sulfamethoxazole—each created by replacing one atom of hydrogen in sulfanilamide by a complex collection of atoms.

discovery frequently provides useful insights into which molecular replacements will prove effective for increasing potency.

β-Lactam Antibiotics Were the Second Great Class of Antibiotics

Following the sulfa drugs, a second discovery of the greatest importance for infectious bacterial disease was made: penicillin. Penicillin is the first of the β-lactam antibiotics, so named because each of these molecules contains a four-member lactam ring (Figure 9.2).

The story of the β-lactam antibiotics began in 1928 with the chance observation by Sir Alexander Fleming that accidental contamination of a bacterial culture plate by the mold *Penicillium notatum* created a clear area around the mold, the consequence of killing of the bacterial cells. The mold created and secreted into the culture medium some agent toxic to bacteria. As that agent diffused through the culture medium, it killed bacterial cells in its path, creating the clear, bacteria-free area. The responsible molecule proved to be penicillin, as demonstrated by Howard Florey and Ernst Chain 11 years later. Isolating and characterizing the active principle made and secreted by *Penicillium* was no easy task for a number of reasons. Translating from the Petri dish to a large-scale manufacturing process was unprecedented at the time. Indeed, the first patient treated (an Oxford policeman) received penicillin from cultures grown in bedpans. So precious was the material, that is was subsequently salvaged from his urine, which led to the rumor at the time (1941): "A remarkable substance [is being investigated, and is] grown in bedpans and purified by passage through the Oxford Constabulary."

Penicillin is by no means the only molecule made by this mold and secreted into the culture medium. Thus, it was necessary to separate and purify penicillin away from a family of other molecules. Moreover, penicillin is a somewhat fragile molecule, not the most stable molecular construct of nature. The highly sophisticated technologies for isolation, purification, and characterization of molecules available to chemists today were unavailable at the time. Success with penicillin was enormously notable. Wit, energy, and determination substituted for technology, and remain highly effective human traits today in drug discovery. For their pioneering discovery of the parent molecule of a major class of antibiotics, Fleming, Chain, and Florey shared the 1945 Nobel Prize in Physiology or Medicine.

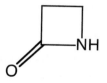

FIGURE 9.2 A β-Lactam, the Key Structure in an Important Class of Antibiotics.

Penicillin is a member of a structural class termed *penams*. A great many penams have subsequently been synthesized and tested, and many have found clinical use, including penicillins G and V, amoxicillin (Trimox), ampicillin (Pen A), methicillin (Azapen), oxacillin (Bactocill), cloxacillin (Cloxapen), carbenicillin (Carbapen), and piperacillin (Zosyn), among others. Like the sulfonamides, these molecules are a structural variation on a common theme. Novel penicillins have improved breadth of antibacterial action; have improved tissue distribution; are effective against organisms resistant to earlier agents; have improved safety profiles, including freedom from allergic reactions to earlier penicillins; and have reduced frequency of dosing. Penicillins include both injectable and oral antibiotics.

Of particular note is the combination of amoxicillin and clavulanic acid. Clavulanic acid was discovered by British scientists working at the drug company Beecham around 1974/1975. After several attempts, Beecham finally filed for U.S. patent protection in 1981; U.S. patent protection was awarded in 1985. Clavulanic acid is a potent inhibitor of enzymes that degrade amoxicillin and many other β-lactam antibiotics. This combination, marketed as Augmentin, increases the efficacy of amoxicillin against organisms that would be otherwise resistant to it. It is among the most widely used antibiotics in the United States. Augmentin as a combination product provides an important precedent for our main interest—Primaxin.

Antibiotics based on a related structural core, termed the *cephems*, are also highly useful. These agents are the cephalosporins and cephamycins, and they include such important antibiotics as cephalothin (Keflin), cefazolin (Ancef), cephalexin (Keflex), cefamandole (Cefam), cefuroxime (Zinacef), cefeclor (Ceclor), and cefoxitin (Mefoxin).

Thienamycin Provided an Opportunity Decorated with Problems

With the exception of the sulfa drugs—products of the chemistry laboratory—all antibiotics in common use at the time of discovery of Primaxin were either natural products or derived from natural products. In addition to the penicillins and cephalosporins, these medicines included the aminoglycosides (for example, streptomycin, neomycin, tobramycin), tetracyclines (for instance, Aureomycin, Terramycin), and macrolide antibiotics (for example, erythromycin, Zithromax, Biaxin). The quinolone antibiotics—products of chemistry, not nature—came along later (for example, Cipro, Floxin, Floxacin, Levaquin).

So it should come as no surprise that the Merck effort to discover novel antibiotics relied on screening natural products—specifically, the fermentation broths of bacteria and fungi. Merck had made one major contribution to the family of β-lactam antibiotics earlier—Mefoxin, a cephamycin derivative—launched in 1978. Mefoxin quickly became the number one hospital-based antibiotic.

Here is a key insight into success in drug discovery: the nature of assays. The biomedical literature is full of examples of important discoveries made on the basis of the quality of an assay, and two examples follow.

During the 1960s, Eugene Delaney, a Merck scientist, had devised a clever assay to search for antibiotics that inhibit bacterial cell wall synthesis. Briefly, growing bacteria are exposed to test substances. Those that are active inhibitors of cell wall synthesis cause the bacteria to form spheroplasts—basically, bacteria without cell walls. Without the structural stability of intact cell walls, the bacteria round up into these nearly spherical structures. The result is detected easily with the naked eye. The spheroplast assay is useful in searching for antibiotics active against Gram-negative bacteria. Antibiotics active in this assay are expected to be bactericidal rather than bacteriostatic. In other words, they kill bacteria rather than prevent their continued growth. Using this assay, Merck discovered the antibiotics fosfomycin and cephamycin C.

So successful was the Merck program in detecting activities against the Gram-negative organisms, distinguishing novel entities from the already discovered ones became problematic. New methods of screening always lead to a burst of new discoveries. So it was in this case: a detection technique based on the Gram-positive species *Staphylococcus aureus* led to the detection of thienamycin. Jean Kahan developed this assay and had been invited to accompany her husband and fellow scientist, Fred (whom we shall meet frequently in what follows), on a trip to the Spanish laboratory Compañia Española de la Penicilina y los Antibioticos (CEPA), which conducted Merck's screens. Rather than go empty-handed, she rapidly devised a technique based on a shift in color of fluorescence that occurs when certain dyes bind to cell walls that have been disrupted by the action of β-lactam antibiotics. A border of green fluorescence resulted, which the Kahans abbreviated as FLECO, their acronym for "fluorescence of ecdysial origin." (The term *ecdysiast* was featured prominently as a come-on to Amsterdam by KLM at the time.) Thienamycin was discovered 3 months later, on September 24, 1973 (9/24), and acquired, coincidentally and fittingly, the serial number 924A.

Antibiotics that act to prevent bacterial cell wall synthesis target specifically the biosynthesis of peptidoglycans, which are key constituents of the bacterial cell wall. Related structures do not occur in cells of higher organisms that, after all, do not have cell walls. Inhibition of biosynthesis of structures that are unique to bacteria is an attractive and well-validated strategy for antibiotic discovery. Penicillins and cephalosporins work this way.

The Gram-staining protocol is a means of distinguishing between the two classes of bacteria mentioned earlier: Gram positive and Gram negative. The bacteria are stained with the dye crystal violet and then washed. Gram-positive bacteria retain the stain whereas Gram-negative ones do not. Gram-negative bacteria are then stained with safranin, which turns them pink or red for easy detection. Differences in cell wall structure account for the staining dichotomy. Gram-negative organisms have an outer cell wall structure that is missing in Gram-positive ones. However, both have peptidoglycan structures in their cell walls. Typical Gram-positive bacteria include the genera *Actinomyces, Bacillus, Clostridium, Enterococcus, Lactobacillus, Listeria, Mycoplasma, Streptococcus*, and *Staphylococcus*. Gram-negative bacteria include the genera *Acinetobacter, Escherichia, Klebsiella, Pseudomonas, Vibrio, Haemophilus*, and *Yersinia*. In general, treatment of Gram-negative infections is more difficult than treatment of Gram-positive ones. Many Gram-negative bacteria are resistant to a range of antibiotics. Ideally, one would like to have a broad-spectrum antibiotic effective against both classes of bacteria.

During the course of screening soil microorganisms, Merck scientists discovered a fermentation broth of particular interest. It had broad-spectrum antibiotic activity. Like penicillin G, the antibiotic responsible for the activity (thienamycin) was very unstable. Efforts to concentrate it led to inactivation. The Kahans, both biochemists, attributed their ability to endure such instability to their experience with enzymes, molecules in another class but with the instability liability. Indeed, this fermentation broth would not have been investigated further were it not for an exceptional property found in concentrated extracts: it was active against *Pseudomonas aeruginosa*, a problematic Gram-negative pathogen. *P. aeruginosa* infections can cause pneumonia, urinary tract infections, and bacteremia. It is the pathogen isolated most frequently from patients who have been hospitalized for several days. These infections are complicated to treat and can be life-threatening. So the activity against this pathogen plus activity against pathogenic Gram-positive bacteria was highly attractive. Beyond that, the broth proved to be active against bacteria that were resistant to a spectrum of penicillins and cephalosporins.[2] When subsequently, Helmut Kropp (a member of the Kahan group) demonstrated that this test tube activity against diverse microorganisms extended to the protection of mice infected experimentally by these species, the full enterprise of chemists and biologists were enlisted in pursuit of this valuable agent.

An early question was: what is the organism, isolated from the soil of New Jersey, that produces this antibiotic activity? It was a *Streptomyces* species, but which one? Based on its morphology, the absence of a melanoid pigment, and its orchid pigmentation, Merck microbiologists established that the organism was an unknown species of *Streptomyces* and elected to name it *Streptomyces cattleya*. The species name derives from the botanical orchid genus *cattleya* and reflects the color of the bacteria (Figure 9.3). To understand the properties of the antibiotic activity observed in the fermentation broth, it was necessary to purify the active material from the soup. And that is where the problems began.

THE ISOLATION OF THIENAMYCIN WAS HIGHLY PROBLEMATIC

There were three reasons why isolating the antibiotic, subsequently named *thienamycin* on the basis of its structure, from the fermentation broth was so difficult: (1) *S. cattleya* produces very little thienamycin, (2) it is insoluble in the solvents usually used to isolate natural products from fermentation broths, and (3) as noted earlier, it is unstable.[3] Let's look at these issues one at a time.

Efforts to isolate and purify thienamycin from the *S. cattleya* fermentation broth were initiated by growing the organism in a 5,670-L (about 1,600-gal) stainless steel fermentation tank. To simplify a bit, let's say you start with about 5,000 L of broth. This is clearly an industrial size effort, not something that you can do in your basement or garage. Fermentation yields of thienamycin varied some from run to run, but typically the yield of thienamycin was about 2 µg/mL (or 2×10^{-6} g/mL or 0.000002 g/mL), which is not much. Doing a bit of arithmetic will reveal that 5,000 L of fermentation broth would contain about 10 g (about one-third of an ounce) of thienamycin. Most of it would be lost during the course of purification efforts. So the poor yield of the desired product in the fermentation broth was problem 1.

Most natural products of biomedical interest are sparingly soluble in water but quite soluble in a number of organic solvents such as methanol. The first step in purification frequently involves the extraction of the active material from the fermentation broth into an organic solvent. From there, powerful methods of purification are available to be used. Thienamycin proved to be quite soluble in water but sparingly soluble in the usual organic solvents. This solubility issue was a major inconvenience. Many of the purification procedures usually used were unavailable to Merck scientists, who were trying to sort out the bit of thienamycin from the ocean of other stuff in the fermentation broth. They had to rely on a sequence of inefficient steps to get the job done. This was problem 2.

Last, and most crucially, thienamycin is an unstable molecule. It self-inactivates. The instability takes a number of forms. When solutions of the antibiotic are frozen, thienamycin self-inactivates. So isolation chemists would finish their day's work, put samples in the freezer, resume work the next day, thaw the samples, only to find that they had lost most of the work from the previous day. In water, thienamycin is unstable under acidic conditions (low pH), and under alkaline conditions (high pH). So it was necessary to maintain thienamycin solutions near neutrality, pH 7 or thereabouts, throughout the purification procedure. In addition, thienamycin is increasingly unstable as its concentration is increased; it reacts with itself (self-inactivation).[4] As the purification process succeeds and increasingly concentrated preparations are obtained, the instability problem gets worse—problem 3.

These were formidable problems to overcome, and at least one leader argued to forget the whole thing—too difficult. Fred and Jean Kahan decided to start a laboratory isolation process and work nonstop until they had a preparation suitable for structure determination. This required 3 to 4 days of continuous work, with Fred sleeping a few hours from time to time in Jerry Birnbaum's (Fred's boss and a Merck Research vice-president) conference room. Jean retreated to their home for her bits of sleep. Fred and Jean succeeded. A nearly pure sample was obtained and turned over to the structure determination scientists. Jerry Birnbaum convened a celebration in a large conference room where success was toasted with a large bottle of Yago sangria in recognition of the Spanish screening laboratory where the organism had been isolated from a New Jersey soil sample. The first peril of the Primaxin project had been overcome by a heroic effort under difficult conditions.

The next issue was: what was this stuff? Merck needed to determine the structure of the thienamycin molecule.

THE THIENAMYCIN STRUCTURE PROVED TO BE A TOUGH NUT TO CRACK

In addition to thienamycin, the S. *cattleya* fermentation broth contained two known β-lactams, penicillin N and cephamycin C. The presence of these β-lactams of traditional structure was misleading to those scientists attempting to deduce the structure of thienamycin. Who would have thought that the same organism would produce both β-lactam antibiotics of the usual sort (penams and cephems) as well as one of an unheard of sort? Nature threw a curveball.

Most of the work on the structure of thienamycin had to be done on derivatives of this molecule because of its instability.[5] What was learned from the derivatives had to be related back to the parent compound, thienamycin. There is no easy way to describe the process leading to the identification of the structure of thienamycin. Most of the techniques of modern structural chemistry were used and they are highly technical, including infrared spectroscopy, proton and carbon-13 nuclear magnetic resonance spectroscopy, X-ray fluorescence, and X-ray crystallography. Suffice it to say that Merck scientists did solve the structure—all its details—in 1978. The formal publication listed 14 authors and all made useful contributions.[6] Unraveling the structure of the unstable thienamycin molecule is a milestone in small-molecule structure determination. Success here depended on increasingly powerful tools of physical organic chemistry that permitted detailed

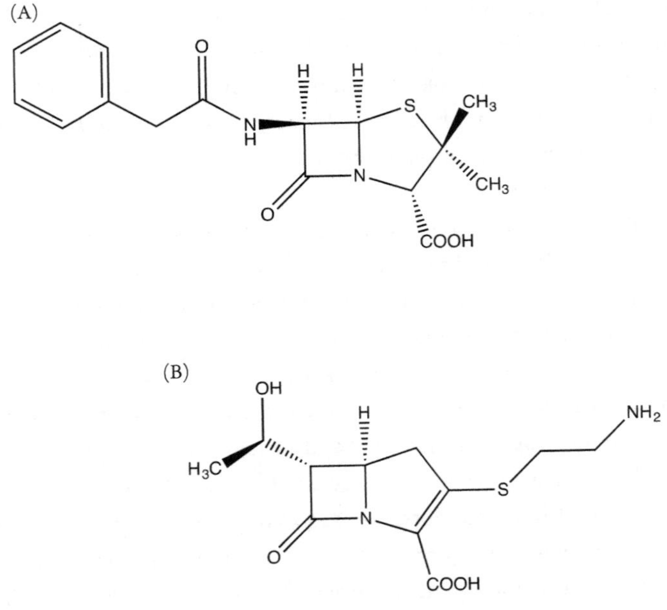

FIGURE 9.4 (A) Penicillin G. (B) Thienamycin.

work on submilligram quantities of material. This work rescued the Primaxin effort from the second peril. The structure, once known, prompted pursuit of a chemical synthesis of thienamycin that could bypass its cumbersome isolation from broths. Ultimately, the commercialization of imipenem made use of these purely chemical routes.

Although I have tried to eliminate chemical structures wherever possible, I am going to include two here: penicillin G and thienamycin (Figure 9.4).[7]

Penicillin G is a typical penicillin. It has the basic structural penicillin skeleton. Almost all β-lactam antibiotics have two fused rings. Penicillin G has a sulfur atom, S, in one of those rings. In marked contrast, thienamycin has the sulfur atom on a side chain attached to the ring system. Thienamycin was the "first biogenetically novel fused β-lactam antibiotic since the discovery of penicillin." There is a second critical structural difference. Penicillin G has an amino (note the nitrogen atom) side chain structure on the left side of the molecule. In contrast, thienamycin has an alkyl (note the absence of the nitrogen atom) side chain there. The dotted line (chemical bond) attaching the side chain to the nucleus of the molecules has further significance. In thienamycin, this side chain points from the nucleus in the opposite direction found in penicillin. It is this "trans" conformation that gives thienamycin its ability to act on strains expressing penicillinases or cephalosporinases, both examples of β-lactamases. This difference is, therefore, key to the resistance of thienamycin to enzymes (β-lactamases) that degrade most β-lactam antibiotics.

THE NEXT MAIN ISSUE WAS HOW TO MAKE THIENAMYCIN INTO A DRUG

Having reasonable amounts of essentially pure thienamycin in hand permitted Merck biologists to define its spectrum of antibiotic activity. More about this follows later, but, for the moment, understand that thienamycin lived up to expectations: potent; broad spectrum, including *Pseudomonas*; and effective against organisms resistant to other β-lactam antibiotics. The problem at this point was that the instability of thienamycin

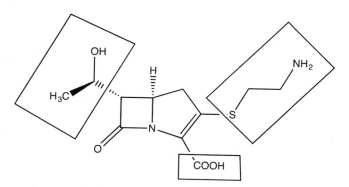

FIGURE 9.5 The Strategy for Chemical Modification of the Thienamycin Molecule. The three chemical groups enclosed in boxes were modified systematically in an effort to gain stability without loss of the highly attractive antibiotic properties of thienamycin itself.

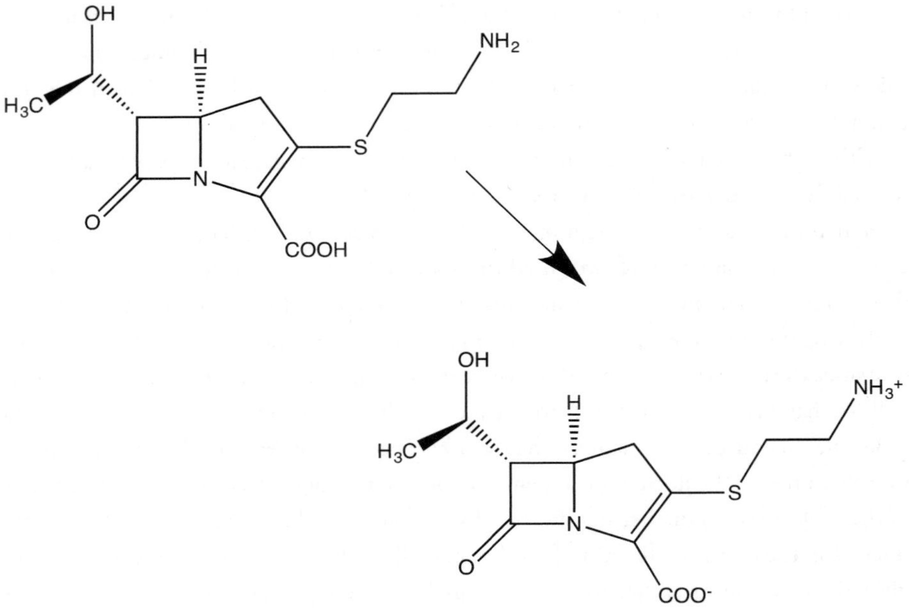

FIGURE 9.6 Thienamycin Can Exist in Two Forms. When the uncharged structure of thienamycin (upper left) is present in aqueous solution near pH 7 (neutrality), a proton is transferred from an oxygen atom of the COOH group to the nitrogen atom of the NH₂ group with the formation of a structure bearing two charges, one positive and one negative (lower right).

meant that it was unsuited to be a clinically useful drug. Perhaps all the work had been for nothing. The Primaxin project faced another peril.

The questions facing Merck scientists were: can the structure of thienamycin be altered in a way that retains its essential antibiotic features but overcomes the stability issue? Is there such a molecule? Or will modifications to the structure to provide stability also compromise the antibiotic properties that generated intense interest in the first place?

Merck chemists under the leadership of Burt Christensen set about making thienamycin analogs—variations on the theme of the thienamycin structure—and biologists under the leadership of Jerry Birnbaum set about testing them. The thienamycin required for this work was supplied through the methodology worked out by Fred and Jean Kahan and the efforts of Merck's isolation chemists.

Merck chemists explored variations in three parts of the thienamycin molecule, and they are boxed in Figure 9.5.

A great many analogs were made and tested without success. The structural changes provided stability but lost the *Pseudomonas* activity, a key aspect of the antibiotic profile of thienamycin.

The key insight that led to success was the recognition that a positive charge on the thienamycin side chain was indispensable both for its highly desirable activity against *Pseudomonas* and also for some degree of stability to metabolism. The thienamycin

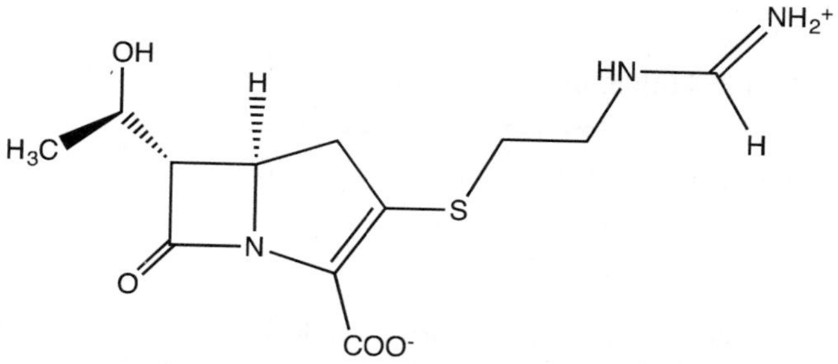

FIGURE 9.7 Imipenem. Note the modest but critical structural changes compared with thienamycin. These provided stability without compromise to antibiotic efficacy.

structure in Figure 9.4B is drawn in the usual manner of depicting a chemical structure. However, thienamycin in aqueous solution at physiological pH bears both a positive charge and a negative one, the consequence of the transfer of a single proton (Figure 9.6).

So, when thienamycin acts as an antibiotic, the sulfur-containing side chain bears a positive charge. It turns out that it was essential to retain this positive charge on the side chain in analogs while maintaining the remainder of the structure as it is in thienamycin itself.

A class of compounds known as *amidines* meets this need. The simplest of these is N-formidoyl thienamycin (Figure 9.7), first made by Tom Miller, a process chemist at Merck. Merck gave it the code name MK0787. Later, it was given the generic name *imipenem* and I use imipenem in the text that follows.

This molecule differs from thienamycin by the addition of a four-atom fragment to the side chain. This subtle change provided a molecule with stability consistent with being a marketed drug and actually improved its antibiotic properties. Success! This triumph of medicinal chemistry rescued the Primaxin project from its third peril.

A SUMMARY OF THE ANTIBIOTIC PROPERTIES OF THIENAMYCIN AND IMIPENEM

To get started, have a look at the bit of data collected in Table 9.1.[8] Values of the minimal inhibitory concentration (MIC) for several antibiotics for two strains of *Enterobacter cloacae*, a Gram-negative pathogen, are provided. These MIC values were obtained from an *in vitro* assay; no living animals were involved. Potency studies against bacterial infections in mice came a bit later. The smaller the value of MIC, the more potent the antibiotic. From the data for strain for 2647 (Table 9.1), you can see that imipenem is about twice as potent as thienamycin and far more potent against this pathogen than three other β-lactam antibiotics—cefoxitin, cephalothin, and carbenicillin—as well as gentamicin, an aminoglycoside antibiotic. The inclusion of gentamicin here is revealing. At the time, gentamicin was the gold standard of potency against *Pseudomonas*. Note that imipenem is at least nine times more potent here than gentamicin.

TABLE 9.1

Values of the Minimal Inhibitory Concentration (in micrograms per milliliter) for Several Antibiotics for Two Strains of *Enterobacter cloacae*

Strain	Imipenem	THM	CFX	CEF	CAB	GEN
2647	0.16	0.32	12.5	12.5	20.0	1.3
2646[a]	0.63	0.63	>100	>100	20.0	0.63[b]

[a]Strain 2646 encodes a β-lactamase, which is lacking in strain 2647; the presence of the β-lactamase makes this strain resistant to many β-lactam antibiotics. [b]Gentamicin is not a β-lactam antibiotic and therefore remains potent against the 2646 strain that encodes a β-lactamase. CAB, carbenicillin; CEF, cephalothin; CFX, cefoxitin; GEN, gentamicin; THM, thienamycin.
Source: Data are from Kropp, H., J. G. Sundelof, J. S. Kahan, F. M. Kahan, and J. Birnbaum. 1980. MK0787 (N-formimidoyl thienamycin): Evaluation of in vitro and in vivo activities. *Antimicrobial Agents and Chemotherapy* 17: 993–1000.

Now have a look at the data for strain 2646. This strain has a β-lactamase, an enzyme that degrades many β-lactam antibiotics, providing it with resistance. Imipenem and thienamycin remain potent against this strain whereas cefoxitin and cephalothin have lost their effectiveness. These antibiotics have a β-lactam ring that is susceptible to degradation by the β-lactamase. Gentamicin retains its potency as expected, because it is not a β-lactam antibiotic and the β-lactamase cannot act on it.

This pattern plays out throughout vastly more expanded experiments.[9] Imipenem proved to be about twice as potent as thienamycin against bacterial species resistant to most β-lactam antibiotics, including *Pseudomonas*, *Serratia*, *Enterobacter*, *Enterococcus*, and *Bacteroides*. Imipenem was also highly effective against antibiotic-resistant clinical isolates of *Pseudomonas aeruginosa*, and *Proteus*, *Enterococcus*, and *Serratia* species.

The *in vitro* studies were followed up with tests of imipenem and thienamycin in infected mice with consistent results. Here, too, imipenem was more potent than thienamycin and far more potent than other β-lactam antibiotics. Imipenem was moved forward into preclinical development, which is where the next problem popped up.

IMIPENEM IS METABOLIZED IN THE KIDNEY

In drug development, there are two important related questions: (1) How is the drug cleared from the human body? (2) What is the route of excretion? These questions are, as are basically all drug development questions, first addressed in laboratory animals. Thienamycin and imipenem are molecules of modest size as drugs go. One would expect them to be cleared by the kidney and show up in the urine. Larger drugs, those with higher molecular weights, are usually cleared by the liver and appear in the feces.

Studies in several species—mice, rats, rabbits, dogs, Rhesus monkeys, and chimpanzees—revealed that urinary recovery of both thienamycin and imipenem was low.[10] In the most extreme case—dogs—only 5 percent of an administered dose of thienamycin was recovered in the urine. These findings strongly suggested that metabolism was a major route of clearance of thienamycin and imipenem in these laboratory animals. In other words, these antibiotics were acted on by one or more enzymes and converted into other molecules that were then excreted. Metabolism by the kidneys can be limited by ligation of the renal arteries. When Helmut Kropp ligated these arteries in rats and rabbits, satisfactory blood levels could be maintained; the duration of thienamycin and imipenem in plasma was prolonged. This strongly suggested that kidney metabolism was an important route of clearance of these antibiotics. In fact, 92 percent of imipenem was cleared by kidney metabolism in rats and rabbits, which is unusual—most drug metabolism occurs in the liver.

On the positive side, kidney metabolism of imipenem did not compromise its effectiveness against systemic infections in animals. On the negative side, this metabolism limited the levels of antibiotic activity in the urinary tract and therefore compromised the effectiveness of imipenem in urinary tract infections. Beyond that, the accumulation of metabolites of imipenem in the kidney might pose a risk of kidney toxicity. In studies of high doses of imipenem alone in susceptible animal species, proximal renal tubular necrosis was observed. Put simply, under extreme conditions, imipenem induced kidney damage—another potential peril for the Primaxin project. Whether this was the result of metabolite accumulation was not clear. However, metabolite accumulation as the basis for potential kidney toxicity was plausible, and if one could eliminate a threat of toxicity, it was worthwhile to do so. So it was deemed important to attempt to limit the extent of kidney metabolism of imipenem.

The first step in limiting kidney metabolism of imipenem was to identify the enzyme or enzymes in the kidney responsible for its metabolism. Helmut Kropp and Fred Kahan identified an enzyme known as *dehydropeptidase I* (DHP-I) as being responsible for at least the bulk of kidney metabolism of imipenem.[11] DHP-I is a zinc-metallopeptidase in the same family as carboxypeptidase A, which we met back in chapter 7. DHP-I acts on imipenem to open the β-lactam ring, obliterating its antibiotic activity.

DHP-I had been described in the biochemistry literature but had not received much attention from enzymologists. Its physiological function was not clear. Aside from its inactivation of imipenem and thienamycin in the kidney, there had not been a whole lot of reason to get excited about this enzyme. No effective inhibitors of DHP-I had been described. To augment the activity of imipenem in the urinary tract and perhaps to limit kidney toxicity, Merck needed a safe and effective inhibitor of DHP-I compatible with imipenem. This need required the initiation of an entirely new drug discovery and development effort, which Merck elected to undertake.

CILASTATIN IS AN EFFECTIVE INHIBITOR OF DHP-I

Back in the 1940s, a Merck scientist had deposited a compound known as benzoylamino-acrylic acid (BAA) in the Merck compound collection. Fred Kahan recognized the structural similarity between this molecule and thienamycin and found it to inhibit DHP-I. This discovery launched, under the leadership of Ed Rogers, a program of systematic

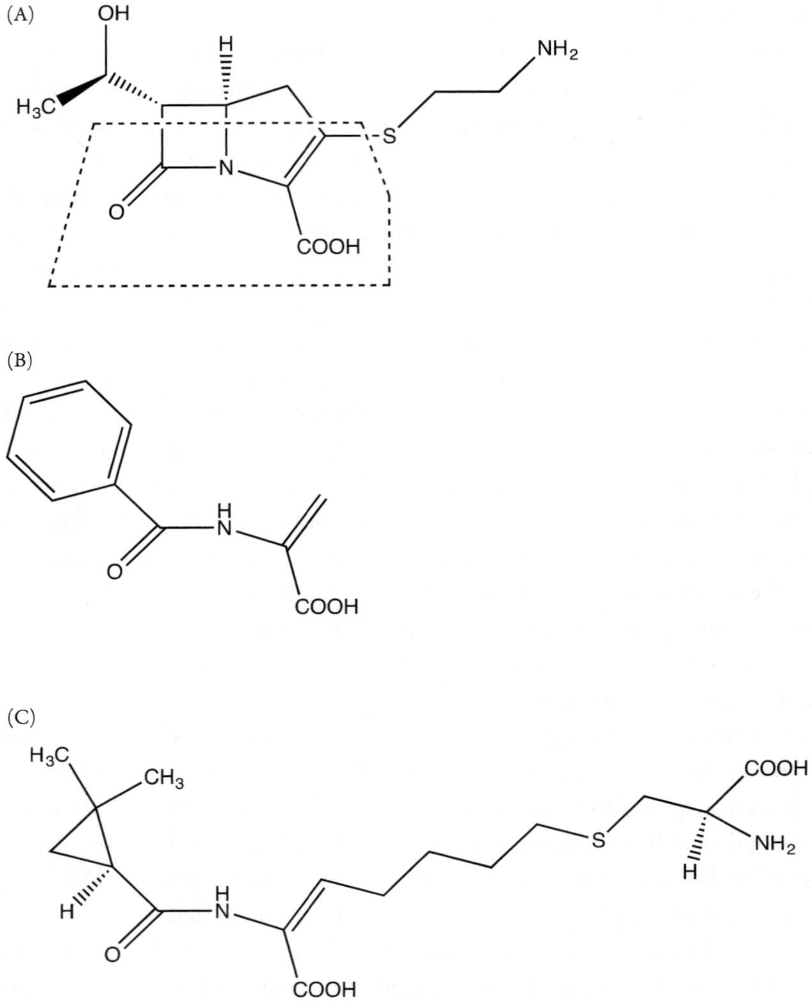

FIGURE 9.8 (A) Thienamycin. (B) Benzoylaminoacrylic acid. (C) Cilastatin. Note that the collection of atoms enclosed by the dotted line in thienamycin is also present in benzoylaminoacrylic acid and cilastatin. This collection of atoms is recognized by DHP-I and is, in part at least, responsible for its action of thienamycin and its inhibition by benzoylaminoacrylic acid and cilastatin.

chemical modification resulting in cilastatin, the partner of imipenem in Primaxin. I make an attempt to illustrate Kahan's insight in Figure 9.8.

Figure 9.8 illustrates the fundamental similarity of structure between imipenem, the starting point of the DHP-1 inhibitor program—BAA (termed a *lead*)—and the final drug, cilastatin. The central point is that the collection of atoms in thienamycin enclosed in the dashed box is also present in BAA. This collection of atoms is recognized by

DHP-I and includes the site of attack by this enzyme. These facts established BAA as a structural foundation for inhibitor discovery.

Cilastatin was the culmination of a drug discovery effort on the part of just four chemists led by Ed Rogers. A successful drug discovery effort usually requires 20 or more chemists working far longer than Ed and his three assistants.

When cilastatin was coadministered with imipenem in a 1:1 ratio, dramatically higher urinary recoveries of imipenem were found, consistent with therapeutic activity for urinary tract infections. In addition, the kidney damage observed in sensitive mammalian species at high doses of imipenem alone was abolished by coadministration of cilastatin. Prevention of nephrotoxicity was the deciding factor in going ahead with imipenem/cilastatin rather than imipenem alone.

There is an amazing side note to this story: the mechanism of protection from the nephrotoxicity of thienamycin by cilastatin is distinct from the inhibition of DHP-I. Cilastatin protects against the nephrotoxic effect of thienamycin simply by excluding the drug from the renal tubular epithelial cell. There are two key experiments that establish this most surprising conclusion. First, the mirror-image molecule of cilastatin has no inhibitory activity against DHP-I, but it is as effective as cilastatin itself in protecting against kidney damage. This is a very clear demonstration that DHP-I inhibition has nothing to do with kidney protection by cilastatin. Second, cilastatin also protects rabbits from the nephrotoxic effects of the classic nephrotoxic cephalosporin cephaloridine, which is not metabolized by DHP-I. Cilastatin protects thienamycin from metabolism and protects the kidney from thienamycin-induced damage, but does so by distinct mechanisms.

The imipenem/cilastatin combination was entered into clinical trials and later became known by its marketed name Primaxin, which I use to refer to the combination in the following sections.

CLINICAL TRIALS OF PRIMAXIN GO WELL

One of the nice things about developing antibiotics is that the results of studies in animal models of infection used to evaluate efficacy usually translate to people. In other words, what works in animal infections usually works in human infections. This is not true with all therapeutic classes. There is no end of drugs that will cure cancer in laboratory rats and mice, but the great majority of them are ineffective in human cancers. Molecules that show promising behavior in animal models of CNS diseases frequently fail in human clinical tests. Drug discovery is not like rolling dice, but it veers in that direction.

Primaxin delivered on its promise in numerous clinical studies: effective and safe. It is indicated for lower respiratory tract infections including pneumonia and bronchitis; intra-abdominal infections such as appendicitis and peritonitis; for skin and soft tissue infections such as abscesses, cellulitis, and wound infections; urinary tract infections; and gynecological infections. There was one final peril.

YOU CANNOT SELL WHAT YOU DO NOT HAVE

While everything else had been going on—isolation of thienamycin, structure determination, discovery of imipenem, kidney metabolism, identification of DHP-I as responsible for kidney metabolism, discovery of cilastatin, evaluation of the combination drug—there had been intensive work to solve a different problem: a source of thienamycin.

As I noted earlier, *S. cattleya* makes very little thienamycin under the culture conditions in which it was discovered. This situation simply would not do as a source for marketed material. There are no fermenters in the world big enough to create enough thienamycin for conversion to imipenem and to serve market needs. Beyond this, the separation process on the necessary scale would have been hopeless. Think about isolating ton quantities of thienamycin out of fermenters the size of Lake Erie.

Merck took two avenues to meet this need. The first was the more typical one when dealing with a natural product: find conditions in which the yield of the desired product proves adequate to meet the need. There are two ways to try to do this. They are by no means mutually exclusive and have the potential for synergy. The first is to take the microorganism as one finds it and explore culture conditions to maximize product yield. In doing so, there are a lot of variables to consider, such as temperature, oxygenation, rate of agitation, carbon source, minerals and other nutrients, and so on. By altering the variables, it usually proves possible for developmental microbiologists to improve yields of the desired product markedly. Despite an intensive effort, fermentation conditions that would encourage *S. cattleya* to produce adequate amounts of thienamycin were not found.

The second thing that one can try to improve product yield is to mutate the microorganism. One hopes to find a mutant *S. cattleya* that produces a dramatically improved yield of thienamycin. Of course, when one has a promising mutant, fermentation conditions can be optimized as noted earlier. Merck scientists made a heroic effort to find such a mutant without success.

While all the fermentation work was ongoing to improve yields of thienamycin, Merck reluctantly explored the other possibility as well—total synthesis of thienamycin followed by conversion to imipenem. This option was unattractive because thienamycin is a difficult molecule to synthesize in the laboratory. Nonetheless, Burt Christensen's chemistry group in Rahway, New Jersey, did complete the total synthesis of thienamycin, which was a major achievement. Without question, it was the most complex total synthesis of an antibiotic at that time. A group known as Process Chemistry under the leadership of Seemon Pines took on the task of finding a way to scale up the complex synthetic process to make large quantities of thienamycin and imipenem—no small challenge.

The challenge was met with ingenuity. Carlos Rosas, a process chemist, located manufacturing facilities for early steps for imipenem—a perfume factory in Switzerland. He also found manufacturing facilities for the early steps for cilastatin— a pyrethrin (a pesticide) factory in Japan. Without these facilities, the capital costs for Primaxin manufacture would have been prohibitive. The final peril to the Primaxin project had been overcome.

The Hallelujah Moment Arrives in 1983

Primaxin was first marketed in Sweden in 1983 under the brand name Tienam. The FDA approved Primaxin for marketing in the United States in 1985. Having overcome the source issue, the product was launched into the domestic marketplace shortly thereafter. Hallelujah! Among infectious disease physicians, Primaxin became known (and is still known) as *gorillamycin*, a testament to its power.

Primaxin is a great success. It was the first carbapenem antibiotic on the market. It was not the last. Carbapenem antibiotics that followed Primaxin into clinical use include meropenem (Merrem), ertapenem (Invanz), and doripenem (Doribax). These carbapenems are resistant to DHP-I and are available as single entities. All have made and continue to make useful contributions to human well-being. The use of carbapenem antibiotics continues to increase.

All antibiotics introduced into clinical practice have at least one thing in common: resistance eventually develops. So it is with the carbapenems. One of the most troublesome pathogens has been dubbed *carbapenem-resistant* Klebsiella pneumoniae. *Klebsiella* is a Gram-negative bacterium that is a cause of infections in health care settings, including hospitals. The infections include pneumonia, meningitis, wound and surgical site infections, and bloodstream infections. These infections are a serious threat to patient safety because carbapenems are the last line of defense against Gram-negative infections resistant to other antibiotics. The mechanism of carbapenem resistance has a familiar ring to it. Resistant organisms have acquired a transmissible genetic element (a plasmid) that encodes enzymes mechanistically similar to DHP-I but distinct from it. These enzymes degrade carbapenems and have been dubbed *carbapenemases* after their function. Unhappily, they are not inhibited by cilastatin. There is a compelling need for a continued flow of novel antibiotics effective against antibiotic-resistant organisms such as carbapenem-resistant *Klebsiella pneumoniae*.

To summarize the stages and perils of the Primaxin project: (1) detection of the antibiotic activity in a fermentation broth, (2) isolation of the unstable activity, (3) determination of the structure of a unique β-lactam, (4) kidney metabolism of imipenem by DHP-I, (5) the discovery of cilastatin, and (6) the total synthesis of imipenem and cilastatin as required for commercial production. Few successful drug discovery and development efforts have been comparably problematic!

For their contribution to the development of Primaxin, Merck awarded their highest measure of esteem, the Director's Scientific Award, to Fred Kahan, Burt Christensen, and Carlos Rosas.

One final note. It seems to me that most drug discovery and development efforts succeed because one person has both the wit and amazing tenacity to work through problem after problem. For Primaxin, that person is Fred Kahan.

It was not a neat event; it is not a "gee whiz" story. No cerebral light bulb flashed the form and function of ivermectin on anybody's mental screen. Nor was the discovery of this new drug the result of an industrial research team's deciding to take existing knowledge at point A and develop it to point B, where it would yield a product to be sold. Nor, yet again, was it a matter of chance, in which some industrial princess of Serendipe sailed to an antiparasitic landing while bound for other therapeutic territory. No, it was a complicated and unglamorous mixture of all of these.[1]

10 Avermectins
MOLECULES OF LIFE BATTLE PARASITES

THESE ARE THE words of Dr. William C. Campbell, hereinafter known as Bill. Bill was Merck's leading authority on parasitic diseases. He played the leading role in the discovery of the greatest antiparasitic drugs in history—ivermectin and abamectin.[2] Having said this, and as Bill has pointed out, this drug discovery story, like all of them, was the result of a team effort involving hundreds of people. When you start giving credit by naming scientists who contributed, it is tough to know when to stop. If you try to name them all, you get a telephone book for a small village and you will still miss somebody. The other extreme is to name nobody, but I have already violated that alternative. So I will mention four scientists at Merck who, in addition to Bill, were the authors of the publication in the prestigious journal *Science* of the article announcing the discovery of ivermectin. They are Mike Fisher, the chemist who led the chemistry effort at Merck focused on the avermectins; Ed Stapley, who led Merck's natural product screening effort; Georg Albers-Schönberg, who headed the group that elucidated the structure of the avermectins; and Ted Jacob, Merck's leader of animal drug metabolism.

Before getting into the story, a word about names—this time about molecules, not scientists. As I relate in this chapter, avermectins are a small family of related molecules. Ivermectin is a chemically modified derivative of one of the avermectins, and abamectin is one of the avermectins.

Bill Campbell is an Irishman and native of Donegal. He took himself to Trinity College in Dublin for his undergraduate work where he did research under the direction of J. Desmond Smyth, a noted parasitologist. Near the end of Bill's undergraduate days at Trinity College, Dr. Arlie Todd of the University of Wisconsin wrote Smyth to ask whether he had any promising students to recommend for graduate study. Smyth recommended Bill to Todd, along with two others. Bill was the one who accepted the opportunity and acquired his PhD there with Todd and Chester Herrick. In a variation on the theme of history repeating itself, at about the time Bill was finishing up at Wisconsin and beginning to look for opportunities in the academic world, Todd received a letter from Ashton Cuckler at Merck in Rahway, New Jersey, asking if he had any promising young people to recommend. Todd recommended Bill, Bill visited Merck, Merck made him an offer, and Bill accepted. Although he still had an academic position on his radar screen, Bill spent all the years until his retirement at Merck. It proved to be a great match, as the story of the avermectins will demonstrate. Based largely on this work, Bill was elected to the National Academy of Sciences of the United States, a highly significant honor and one not frequently bestowed on scientists in industry.

The Merck–Kitasato Collaboration Led to the Discovery of Avermectins

Merck has a long history in the discovery of antiparasitic drugs. Perhaps the most important example before the avermectins is the molecule thiabendazole. Approved for human use in 1967, it was the first anthelmintic in the benzimidazole class. A word about nomenclature. An anthelmintic is a drug that expels parasitic worms from the body. Thiabendazole is used to control infections by roundworms, hookworms, and others. It acts as an inhibitor of the helminth enzyme known as *fumarate reductase*. The product for human use is a chewable tablet marketed as Mintezol. Other anthelmintics in the same class have followed (although not all from Merck), including albendazole (Albenza) and mebendazole (Vermox).

Thiabendazole also has antifungal activity. There is a product for animal use in which thiabendazole is combined with neomycin (an antibiotic) and dexamethasone (anti-inflammatory) for certain fungal and bacterial infections in cats and dogs, and is sold under the brand name Tresaderm. Last, thiabendazole is used to prevent and treat Dutch Elm disease under the brand name Arbotect. Merck had useful relevant experience in animal and plant health, as we shall see.

At the time the avermectin story began, Merck was not focused on finding the next thiabendazole or the next anything. The goal was to discover something totally novel—a game-changing molecule. This is not so simple. Game-changing molecules are few and far between. Merck elected to look for their breakthrough in the area of natural product chemistry—molecules made by living organisms. Merck was already heavily invested in natural product chemistry and had elected to look beyond the borders of the company in their search for the next great molecule. Specifically, Merck had a major collaboration with the Spanish company Compañia Española de la Penicilina y los Antibioticos, known around Merck, for obvious reasons, as CEPA, to screen microbial fermentation broths for interesting biological activities. However, there was no reason to limit Merck's outreach to Spain. Japan was the next outreach target, an entirely reasonable objective given Japan's lengthy experience with fermentation.

Boyd Woodruff, Merck's chief microbiologist, journeyed to Japan to seek out a collaborator in the natural product area with a focus on antibiotics. The antibiotics might be used for human health purposes and perhaps for animal health applications as well. It was known, for example, that antibiotics are growth permittants in cattle. Bacterial infections can limit the rate at which cattle grow, and antibiotics can eliminate or reduce that limit. Not a particularly exciting opportunity in my view, and not where the focus eventually landed.

Boyd met with Toju Hata, president of the Kitasato Institute, and Satoshi Ohmura, one of its leading scientists. There turned out to be an important connection between Merck and the Kitasato Institute. Max Tishler was the legendary leader of Merck Research before his retirement and move to Wesleyan University in Connecticut. Ohmura had spent time in the Tishler laboratory at Wesleyan, so Max knew him well. On this basis, Tishler recommended the Kitasato Institute to Merck and Merck to the Kitasato people. Max Tishler continued to impact Merck long after his retirement from the company.[3]

HERE IS A WORD ABOUT STRATEGY IN NATURAL PRODUCT SCREENING

One downside of natural product screening for molecules having useful biological activity is that you tend to rediscover old molecules; a great many natural products have been discovered over time. Detecting good bioactivity in a biological sample, isolating the active agent, and determining its structure only to find that someone else did it years ago is discouraging. At the time, Merck was running the largest microbial screening effort in the pharmaceutical industry, largely in Spain and secondarily in Rahway. The work was under the direction of Ed Stapley, who reported to Jerry Birnbaum, a vice-president of Merck Research. The effort was plagued by "culture redundancy." If you keep screening cultures of the same organisms, you keep finding the same active molecules.

In an effort to limit the discouragement inherent in natural product screening and to optimize chances for a favorable outcome, there are a couple of things you can do. One is to look for exotic biological samples that might have been overlooked by earlier workers. These samples might include plants collected from truly remote places on Earth, marine organisms scraped up from the deep ocean floor, venoms of poisonous snakes and spiders, soil samples collected from anywhere you can think of—the more exotic the better. The idea is to look where few or none have looked before in the hope of finding something new and useful.

The Kitasato scientists collected soil samples and they did it from multiple places. From each soil sample, they cultured microorganisms on agar medium in plates. They then did the second thing that one can do to try to avoid rediscoveries; they selected the most unusual looking bacterial colonies from all those that grew from a soil sample. The hope was that an unusual colony appearance would translate into a rare or unknown bacterium. Perhaps rare bacteria would produce unknown molecules and these would have useful properties.

This was an attractive opportunity for Merck. After a discussion with Fred Kahan, a leading Merck scientist who we met in chapter 9 (and who we will meet again), Jerry Birnbaum told Ed Stapley that Merck needed to get cultures from Satoshi Ohmura. Negotiations ensued, agreement was eventually reached, and cultures started arriving in Rahway, New Jersey, about 20 at a time.[4]

MERCK CREATED A UNIQUE ASSAY FOR ANTIPARASITIC ACTIVITY

Merck incubated the Kitasato bacterial isolates in a liquid culture medium. As the bacterial population grows, a fermentation broth is formed. Some molecules created by the bacteria are excreted into the culture medium whereas others are retained in the organisms. One gets a rich mixture of everything the bacteria make under the conditions of their growth. The issue is to discover if this soup contains anything of interest. For this, one needs an assay, some way of measuring the presence of what you are looking for. Merck was free to examine these fermentation broths in as many assays running in Rahway as they wished. Only one is of interest to us.

In three of the four previous chapters, the assays for desired biological activity were simple: inhibition of the activity of an isolated enzyme: target-based drug discovery. Test tube assays provided the basis for Merck's discoveries of finasteride, enalapril, lisinopril, lovastatin, and simvastatin. The search for antiparasitic activity in the fermentation broths from the Kitasato bacterial isolates was substantially more complex. The assay was conducted with mice, an example of a phenotypic assay, which is not simple. With assays for enzyme inhibitors, it is now fairly routine to screen thousands of samples a day, usually using robots to do the work. You cannot do these screens involving living organisms. Constructing a useful screen for antiparasitic activity in mice was critical to Merck's success. Here is what they did.

Each mouse was infected with two parasites: a coccidian (*Eimeria*) and a roundworm (*Heligmosomoides*). Merck was interested in both activities, and the dual-infected mice provided the opportunity to get two bites at the apple from each mouse.

To concentrate the fermentation broths and, therefore, the biologically active molecules within them, Merck froze them and dried them under a vacuum. The freeze-dried broths were then added to the feed for the dual-infected animals. Each broth was tested in only one mouse. The animals were permitted to feed for a few days and were then sacrificed. The assay consisted of looking for the absence of parasites in the gut and the absence of worm eggs in the feces. If you are thinking that this is not the kind of work that would appeal to you—sorting through mouse guts and feces looking for evidence of parasites—no surprise. But, this is what Merck scientists did. I doubt that anyone would run this type of assay today, but perhaps they should. This assay was up and running at Merck in Rahway in 1974.

ANTIPARASITIC ACTIVITY WAS DISCOVERED IN THE FERMENTATION BROTHS

Among the fermentation broths derived from the Kitasato bacterial isolates, one was found to be active against roundworms in the dual-infected mouse assay. That mouse was found to be free of roundworms in the gut and roundworm eggs in the feces. Something in the fermentation broth had eliminated all evidence of roundworms from an infected mouse. It was very nearly missed. The mouse in question nearly died before completion of the feeding regimen. This turned out to be the consequence of a toxic molecule in the fermentation broth independent of those showing antiparasitic activity (remember, these formation broths are complex mixtures of molecules). The active fermentation broth was given the Merck code name MA-4680.

The source of the sample yielding the active bacterial isolate was soil taken by Ohmura and Ruiko Oiwa near a Japanese golf course. Who would have thought to look there? The bacterium was found to be a novel species of *Streptomyces* and was named *Streptomyces avermitilis*. The Kitasato scientists continued to work on this organism for many years. In 2003, Satoshi Ohmura and his colleagues reported the complete genomic sequence of this bacterium.

The discovery of *S. avermitilis* among the Kitasato samples is utterly amazing. At least 250,000 microbial cultures, including a great many at CEPA, have been subsequently screened without finding this bacterium again! This discovery is more than mere luck. It depended on knowing what you were looking for in the first place, having the wit to look broadly for soil sources of microorganisms, having the insight to select exotic-looking bacterial cultures for further studies, having a great screen to search for the desired biological activity, and running the screen properly. Clearly, the stage was set for a fortuitous finding—an example of serendipity, not luck.

Having discovered antiroundworm activity in MA-4680, the next question was: what are the active molecules in this fermentation broth?

The Avermectins Are the Active Agents in MA-4680

The active antiparasitic molecules in MA-4680 were isolated and their structures determined by Georg Albers-Schönberg. Note that there was an enormous amount of wit and work that went into the realization of that simple sentence. These molecules are collectively known as the *avermectins*. There are eight of them in MA-4680: two pairs of four.

The structures are closely related, variations on a central structural theme. Revealing a total lack of originality but a religious devotion to regularity, they were named avermectins A1a, A1b, A2a, A2b, B1a, B1b, B2a, and B2b. The molecules are complex and are based on a 16-member ring system highly decorated with chemical groupings, including exotic sugars (Figure 10.1).

These complex structures evidently evolved over time as a means of protection for the bacteria producing them. The structural complexity also provides chemists with major challenges, a form of job security.

Having molecules in hand with promising biological activity, and the avermectins surely qualified, medicinal chemists got to work to determine whether they could modify them in ways that made them better. That is, after all, why pharma houses hire and pay medicinal chemists. They are very good at their work, and they partner with biologists who create and run the assays that reveal whether an improvement has been made (see the process detailed in chapter 5). In general, the chemists create a new molecule; the biologists determine its properties in relevant assays and provide the information to the chemists who use it to design the next molecule. After a bunch of iterations, progress is usually made (although getting to an end point is another matter; progress may or may not be good enough). So it was with the avermectins. Under the leadership of Mike Fisher, a substantial number of avermectin analogs were made. The best one is known by the revealing (to chemists) but unwieldy (to everyone else) name as *22, 23-dihydro-avermectin B1a*. It is also known as *ivermectin* and it is a triumph of pharmaceutical research.[5]

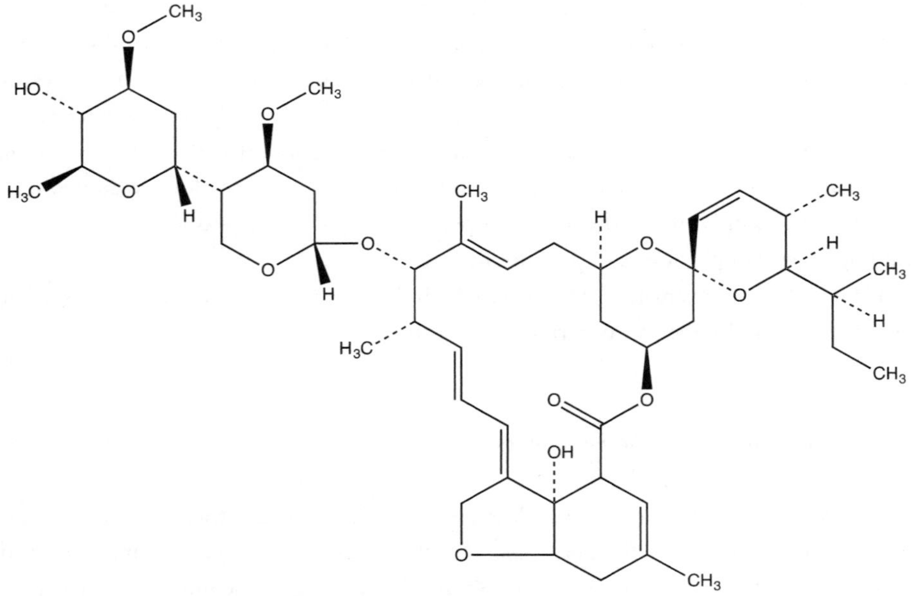

FIGURE 10.1 Avermectin A1a.

THE BIOLOGY OF THE AVERMECTINS IS AMAZING

The discovery of the avermectins generated a tsunami of assays to define the biology of these agents. The assay that led to their discovery demonstrated that they were active against one roundworm: *Heligmosomoides bakeri*, the roundworm with which the mice were infected. How far did this antiroundworm activity extend? Did it extend to other classes of parasites? Merck scientists set out to answer these and other questions. Here is a quick summary of what they learned using ivermectin as the example.

First, ivermectin is active against numerous roundworms (known technically as *nematodes*, which are discussed more later). It is active against benzimidazole-resistant roundworms, such as those resistant to thiabendazole (recall that thiabendazole was an antiparasitic available at the time that ivermectin was discovered), a matter of much interest and suggestive of a novel mechanism of action. Second, ivermectin is active via several routes of administration. It can be given orally, by injection, in feed, and topically (a formulation can be poured on an infected animal). Third, ivermectin is extremely potent, being effective at very small doses. Fourth, ivermectin is well-tolerated by animals (with the exceptions of Collies,[6] Murray Gray cattle, and CF1 mice). Fifth, ivermectin is active against a number of insect parasites. Sixth, ivermectin is active against some acarine parasites (air-breathing arthropods with simple eyes and four pairs of legs, including ticks and mites).

Thus, ivermectin is both an endoparasiticide, killing parasites within an animal such as roundworms, and an ectoparasiticide, killing parasites living on the animal such as ticks, which is quite unusual and really quite neat. This broad spectrum of biological activity is extraordinary. On this basis, Merck created a new division (MSD AgVet) to exploit the potential of the avermectins for animal health and agriculture.

ROUNDWORMS ARE IMPORTANT PARASITES

Roundworms are members of the phylum *Nematoda* and are usually referred to as *nematodes*. We will hang in with the term *roundworms*. There are a great many roundworms; about 30,000 species are known and about half of them are parasitic. Most are small and some are microscopic; there are exceptions, however. Roundworms are everywhere; they are in soil, seawater, and freshwater. They are found in the polar regions and in the tropics. By some estimates, 80 percent of all animals on Earth are roundworms. If you look for a roundworm, you are going to find one.

The iconic roundworm is the free-living organism *Caenorhabitis elegans* (Figure 10.2). This is a favorite organism of biologists, a model organism. There are a few species on Earth that are exceptionally well understood. These species include the fruit fly *Drosophila melanogaster*, yeast, mice, and *C. elegans*. For example, *C. elegans* has exactly 302 neurons in its nervous system and these neurons have been mapped precisely. Much has been learned about nervous systems from studies of this simple roundworm.

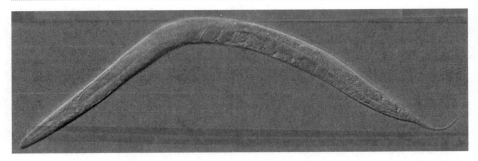

FIGURE 10.2 The Model Organism *Caenorhabitis elegans*. The roundworm is about 1 mm in length and lives in soils in temperate regions. About 95 percent of these worms are hermaphrodites; the remaining ones are males. This worm is one of the most intensively studied, and best understood, organisms on Earth.

The roundworms include the following subcategories: ascarids, filarias, hookworms, pinworms, and whipworms. Among the ascarids is the giant roundworm *Ascaris lumbricoides*, which can be as long as 30 cm (12 in). Ascariasis is the human disease resulting from infection with this roundworm. About one-fourth of the human population is so infected—about 1.7 billion people. Infection may be asymptomatic, although symptoms may include bloody sputum, shortness of breath, stomach pain, and low-grade fever.

Among the filarias is the genus *Dirofilaris*, which causes heartworm in dogs. The point here is simple: roundworms are important parasites of animals and people. Agents that prevent roundworm infection or rid animals and people of roundworms will find abundant use.

AVERMECTINS CAUSE CHLORIDE CHANNELS TO OPEN

Having discovered the profound antiparasitic activity of the avermectins raises two questions. What is the molecular basis of the antiparasitic activity of the avermectins? Why are these molecules safe for higher animals?

Understanding the mechanism of action of the avermectins has been problematic. Here is a simplified but revealing summary. Avermectins are known to act by binding to a subunit of the glutamate receptor (chapter 4) in invertebrate nerve cells. The invertebrate glutamate receptor regulates a chloride ion-specific ion channel. When an avermectin binds to it, the chloride ion channel opens and chloride ions flow through it. The ion flow increases the voltage difference (hyperpolarization) across the neural membrane and inhibits the transmission of an action potential. At the neuromuscular junction, the signal required for muscle contraction is lost and the worm is paralyzed. In sum, avermectins stimulate the glutamate receptor, the chloride ion channel opens, chloride ions flow, membranes become hyperpolarized, nervous system signals are lost, and paralysis results. Worms are swept out; ticks fall off.

The safety of avermectins for vertebrates is the consequence of two factors. First, glutamate receptors in vertebrates regulate channels for positively charged ions such as those for potassium and calcium, not negatively charged ions such as chloride. Second, the glutamate-activated nervous system in vertebrates is mostly localized in the CNS, not in the peripheral nervous system. In vertebrates, the CNS is protected behind the blood–brain barrier. This barrier is the result of very tight junctions between those endothelial cells that line the vasculature of the CNS. The point is to keep molecules that do not belong in the brain out of the brain; there are special mechanisms for allowing access to the brain of the molecules that belong in the brain. The avermectins do not penetrate the blood–brain barrier in vertebrates. Thus, animals and people are protected from the actions of the avermectins. Invertebrates, including a great many parasites, have no such protection. The blood–brain barrier of Collie dogs, Murray Gray cattle, and CF1 mice evidently leak some avermectins, accounting for reduced safety of avermectins in these animals.

Merck Develops Ivermectin for Animal Health Purposes

The biological properties of ivermectin—broad-spectrum endoparasitic and ectoparasitic activity coupled with safety for their hosts—turned on a big development machine at Merck. The goal was to create the data necessary to establish efficacy against numerous parasites in numerous species while demonstrating safety for those species. Approval to market ivermectin in specific species for specific parasites would result. Target species included animals raised for food—cattle, sheep, and hogs, among others—as well as companion animals. Furthermore, optimizing the market potential of ivermectin required approval of these products in numerous countries. Merck focused on numerous products in numerous countries for as many important parasites as reasonably possible.

The discovery of the biological properties of the avermectins created excitement beyond the walls of Merck and Company. Hundreds of publications, mostly from academic laboratories, supplemented the work at Merck.

Commercial use of ivermectin began in 1979 and increased in subsequent years as approvals for new products in new countries were gained. Creating a list of all the products and all their indications for use would be stunningly boring. Here is a quick summary of the products: IVOMEC for cattle, sheep, goats, and swine; EQVALAN for horses; and HEARTGARD 30 for dogs. To amplify a bit, I have collected the products and their indications for cattle in Table 10.1. Have a look and you will get the idea of the spectrum of uses of ivermectin for animal health.

Some applications of ivermectin are interesting as well as useful. Let's think about HEARTGARD for dogs for a moment. The problem is heartworm (*Dirofilaria immitis*) infection. These filarial worms occupy various sites in the heart and cardiac arteries and they are quite large. Killing them is dangerous to the host animal. The dead worms may

TABLE 10.1

Marketed Products and Indications for Ivermectin Use in Cattle (IVOMEC)

Product	Indication
IVOMEC injection	Gastrointestinal nematodes
	Lungworms
	Grubs
	Sucking lice
	Mange mites
	Screwworm fly
	Ticks (aid in control)
IVOMEC oral	Gastrointestinal nematodes
	Lungworms
IVOMEC pour on	Gastrointestinal nematodes
	Lungworms
	Grubs
	Sucking and biting lice
	Mange mites
	Ticks (aid in control)

be swept into the circulation of the lungs, where they can create a pulmonary embolism, ending in death (the vasculature of the lung is very fine and more easily blocked than elsewhere). Happily, ivermectin does not kill adult heartworms. It does kill the worms when they are very young, and this can be done safely. Consequently, HEARTGARD is approved for and used for *prevention* of heartworm infection in dogs, not for *therapy* of established infections.

Ivermectin is far more widely used than its approved uses would suggest. Having the products on the market encourages people to explore alternative utilities. This activity is known as off-label use and is very common in human medicine as well as animal health. The FDA does not regulate how a drug is used after it is approved. Here is a short list of animals in which ivermectin use has been reported: guinea pigs, hamsters, gerbils, rabbits, foxes, wolves, camels, alpacas, llamas, deer, reindeer, impala, bighorn sheep, bison, buffalo, wild boar, elephants, nonhuman primates, birds, fish, and snakes. It has also been used in people, a matter that we get to shortly.

It may come as a surprise that approval of a drug for animal health is at least as complicated as approval of one for human health. This is particularly true for food animals such as cattle, for example. There are three basic issues.

First, there is the issue of drug residues in the animal tissues. After administration of a drug to a food animal, traces of the drug itself or its metabolites may remain in the tissues and may be eliminated only slowly over time. It is necessary either to demonstrate that these residues are no threat to human health (somewhere out there is a guy who eats a pound of beef liver every day and he needs to be protected) or to stop drug administration early enough so that no detectable residue remains. Either way, work is required.

Second, there is an issue of environmental safety. A drug fed to an animal will be excreted either in the urine or feces as the original drug or as its metabolites. So the drug residue enters the environment. Think about a cattle feedlot, for example. If the cattle are treated with IVOMEC, it and its metabolites will be deposited in the feedlot. When it rains, they may be washed into a river or stream. There they may kill a spectrum of invertebrates with environmental consequences that are difficult to predict. So, you need to prove this cannot happen. In the case of ivermectin, drug residues are bound strongly to soil, where they are degraded to biologically inactive materials at rates that depend on exposure to sunlight, temperature, and other factors. Work was required to address this issue.

Third, the central point of animal health products is usually economics, not the protection of the health of the animal for its own sake. There are exceptions. Many people will protect the health of their companion animals—dogs and cats primarily—and their riding horses with as much diligence as they use to protect the health of their children. This commitment to health is not so for animals raised for food. Stock growers are interested in making a living. Animal health products that aid them in doing so will be used. Economics drives the business.

Regulators will surely, and rightly, look differently on an application to market a drug that may protect human life or restore human health from one that will increase farm profits, which tends to raise the bar for approval to market animal health products. So no one should believe that products for animal health, particularly for food animals, are easy to come by and easy to gain marketing approval for. Reality is just the opposite.

AVERMECTINS ARE NOT THE END OF THE STORY

Success generates imitation. The market success of the avermectins encouraged others to search for related molecules with comparable or improved properties. A number of them have reached the marketplace in the years since commercialization of the avermectins. They include doramectin (Dectomax), selamectin (Revolution, Stronghold), milbemycin-oxime (Interceptor, Sentinel), nemadectin, moxidectin (ProHeart 6), and eprinomectin (Eprinex). All are related structurally to the avermectins.

The Avermectins Are Useful in Agriculture

Animals are not the only organisms to benefit from the avermectins; plants do, too. As noted earlier, avermectins are also toxic to a number of arthropods, including insects and mites. Certain insects and mites attack plants and fruits, causing damage. Merck elected to develop avermectin-based products to protect a range of crops, including ornamental plants, cotton, citrus fruits, pears, and a family of vegetables. Without delving into details, let me just say that this work succeeded.

For agricultural uses, Merck screened all eight of the natural avermectins as well as a few hundred semisynthetic derivatives that Mike Fisher and his medicinal chemists created in the laboratory. At the end of the day, chemistry had not improved on nature; avermectin B1a was selected for development.[7] Avermectin B1a is known familiarly as *abamectin* to distinguish it from ivermectin, 22,23-dihydro-avermectin B1a, used for animal health purposes. It is sold under a variety of trade names including, but not limited to, AVID, AGRI-MEK, and VERTI-MEK.

Abamectin is also useful for control of the red imported fire ant, which inflicts painful bites on people and animals. Abamectin for fire ant control is marketed as AFFIRM.

The regulatory issues for agricultural products are similar to those for animal health use. The critical issues are human safety and environmental protection. The human safety issue takes two forms: protection of the consumer and protection of the user or handler. If, for example, you spray an orange orchard with an agent that kills mites, does any of that agent or its metabolites get into the edible portion of the orange? If so, how much? And what are its toxic properties to humans? These and related issues require answers, and getting them requires work. Then there are questions of safety to the individuals who apply the agent to the crop and handle it, be it citrus fruits, cotton, or roses.

Environmental issues are pretty obvious but no less easy to deal with. With human health or animal health products, individuals are dosed. Not so with agricultural products, for which whole fields or entire orchards may be sprayed. There is going to be a lot of the drug spread over a lot of area, which may contain a lot of beneficial organisms as well as pests. Here, too, drug residue may run off into rivers and streams, where additional species will be exposed. These issues can be dealt with, given the right molecule, but work and time are required. Abamectin passed the tests for numerous uses.

Ivermectin Is Useful in Human Medicine

Parasitic infections in humans are a huge health burden. Protozoans, unicellular organisms responsible for malaria for example, worms, and arthropods, including ticks and insects, cause infections. They are particularly prevalent in parts of the Earth where sanitation is poor. Protozoans afflict populations at the bottom of the socioeconomic ladder. Poverty and disease feed on one another.

Although the scope of parasitic diseases of people is broad, I am going to focus first on one: onchocerciasis, commonly known as *river blindness*. Onchocerciasis is most often found in equatorial Africa and the highlands of Mexico and Guatemala. Yemen, Brazil, Venezuela, and Ecuador have lower rates of this infection.

The infective parasite in onchocerciasis is *Onchocerca volvulus*, a filarial worm. We met a worm in this class earlier when we discussed heartworm in dogs. Estimates suggest that about 17 million people have this disease. The disease is transmitted by blackflies and it is most usually seen within a few kilometers of rivers or streams, where blackflies breed. The life cycle of *O. volvulus* is provided in Figure 10.3.

When an infected blackfly takes a human blood meal, infective larvae, microfilariae, may enter the wound and develop into adults that live walled-off in nodules in subcutaneous tissues. These nodules are frequently palpable in infected patients. Female worms release microfilariae, which localize in the skin, conjunctiva, retina, and other tissues. When a blackfly takes a blood meal, it may ingest microfilariae from the skin, completing one life cycle.

The key symptoms of onchocerciasis reflect the localization of the microfilariae in the skin (intense itching, wrinkling, disfiguring lesions, dermatitis) and the eye (conjunctivitis, photophobia, and blindness; Figure 10.4). Onchocerciasis is not a benign disease; it cried out for improved prevention and therapy.

At the time, diethylcarbamazine (DEC) was the drug of choice for onchocerciasis. It leaves something to be desired. Heavily infected patients frequently experience adverse effects, sometimes serious. DEC kills the microfilariae, but the debris may elicit a spectrum of adverse effects including the Mazzotti reaction, which is fever, itching, rapid

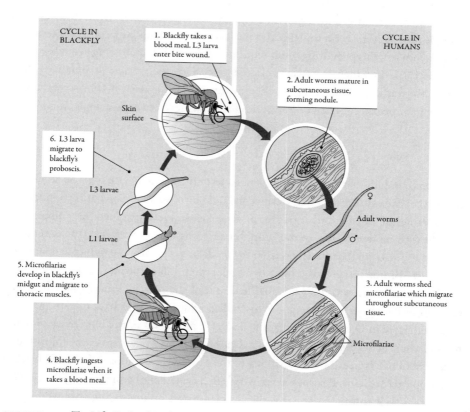

FIGURE 10.3 The Life Cycle of *Onchocerca volvulus*.

heartbeat, low blood pressure, edema, and abdominal pain. DEC is frequently given with a corticosteroid to control the clinical adverse effects.

So how did Merck migrate from the use of ivermectin for animal health to use in humans? Here is a brief chronicle of events.[8]

STUDIES IN HORSES AND CATTLE PROVIDED MOTIVATION

In 1977, Bill Campbell submitted a proposal to Merck management arguing for consideration of ivermectin in human health. Spinning off animal health drugs for use in humans had a precedent at Merck. As noted earlier, the animal health drug thiabendazole (Mintezol) finds use as an antiparasite agent in human medicine.

The data available at that time for use of ivermectin in a filarial disease were not promising. Ivermectin was known not to kill the adult worms and, reasoning from the action of DEC, which does, some concluded that ivermectin had little to offer.

There is a lot to be said for keeping your eyes and your mind open for unexpected opportunities. This is one example.

Merck conducted a trial of ivermectin in horses that were known to have a variety of natural infections of intestinal worms. Horses were treated with ivermectin, and skin samples were examined for *Onchocerca* larva a week to 10 days later. The result was clear. A single dose of ivermectin was effective in clearing the microfilariae of *Onchocerca cervicalis* in horses. This promising result required verification, which came in a subsequent study in cattle in which ivermectin cleared *Onchocerca gutturosa* microfilariae. Thus, ivermectin was known to be highly effective against two species of *Onchocerca* in two mammalian species, suggestive of possible efficacy against *O. volvulus* in people.

IVERMECTIN GETS A CLINICAL TRIAL IN HUMANS

In a memorandum that Bill Campbell sent to Jerry Birnbaum in late 1978, he argued on the basis of the existing data that ivermectin could be of value in preventing skin lesions and blindness in populations at risk for onchocerciasis. He also argued that "discussions be held with representatives of the World Health Organization (WHO) to determine the best approach to the problem—from the medical, political and commercial points of view." Jerry Birnbaum agreed, as did Roy Vagelos, then still president of Merck Research.

Mohammed Aziz was uniquely qualified to lead the Merck clinical trial effort. He came from Bangladesh; and acquired his college education in Calcutta, India, and his medical degree in Dacca, back in Bangladesh. He moved on to the University of Minnesota, where he received a PhD in clinical pathology and from there to the Johns Hopkins School of Hygiene and Public Heath, then on to the London School of Hygiene and Tropical Medicine. He worked for WHO in Sierra Leone before joining Merck. His arrival was well timed. As noted earlier, evidence was accumulating that ivermectin might find use for the treatment of river blindness. One could not have reasonably hoped for a better trained clinician to lead the studies required to test this idea.

In early 1980, Merck management approved a limited phase 1 trial in onchocerciasis. The first clinical trial was sited in Dakar, Senegal, under the direction of Michel Larivière, a professor at the University of Paris and the University of Dakar. There were excellent medical facilities at the University Hospital, abundant expertise in parasitology, and proximity to villages where onchocerciasis was endemic.

This clinical trial was conducted in 1981 and 1982 in 32 infected male villagers from the surrounding territory who had consented to participate. They were selected on the basis of the density of microfilariae in their skin, which was high enough to permit detection by routine skin-snip examination, and low enough to minimize probability of eye invasion and hypersensitivity reactions. Although efficacy was key, patient safety was a critical issue. All trial participants were brought to the hospital for treatment with ivermectin and for subsequent observation.

The clinical trial design included initial doses of 5 to 10 µg ivermectin/kg body weight. These are very low doses, about one 20th of the "no-effect" level determined in toxicity studies in laboratory animals. That is, a dose 20 times greater would not have elicited evidence of toxicity in these animals. Pharma houses generally tread very carefully in first-in-human clinical trials, for good reason. Merck was even more cautious than usual with ivermectin. A central point of this phase 1 trial was patient safety. As confidence of patient safety was gained, doses were scheduled to increase to 30 µg/kg and then to 50 µg/kg.

A cross-over design was selected for the clinical trial (look ahead to Figure 12.13). Half the men were given a single oral dose of ivermectin and half were given a single dose of a placebo. Two weeks later, the treatments were switched; those that had received the placebo now got ivermectin and vice versa. Laboratory assessment and skin-snip examinations were performed at intervals over a 4-week period. The men then returned to their villages where another skin-snip was taken at 6 weeks postdosing.

The results were clear and exciting. There was no reduction in microfilarial density in those patients who had received 5- or 10-µg/kg doses, a significant reduction in microfilarial density at the 30-µg/kg dose, and complete or nearly complete elimination of microfilariae at the 50-µg/kg dose. Side effects were minor and transient.

This was a small but first-rate clinical trial. It was hospital based, well controlled, with careful attention to detail and with extensive clinical support. Aziz wrote up the trial report, published in *The Lancet*, an important British medical journal.[9]

Based on the positive results of the initial clinical trials, a host of others followed to assess the role that ivermectin might play in control of onchocerciasis and other parasitic diseases of humans. Merck turned on a major clinical effort. Phase 2 double-blind, placebo-controlled studies were conducted in Senegal, Mali, Ghana, and Liberia. Phase 3 studies followed. The results were overwhelmingly positive. Ivermectin was approved for use in onchocerciasis in 1986, and the trade name Mectizan was adopted. Onchocerciasis is now treated with single oral doses of 150 µg/kg at intervals between 3 months and 12 months. Gaining control of a miserable and potentially blinding disease with single doses of ivermectin at long intervals in parts of the world where delivery of medical care is problematic is the best outcome that one could have reasonably hoped for. There is a clear possibility that onchocerciasis may one day soon be eradicated, as smallpox has been. The only host for *O. volvulus* is people. As microfilariae of this parasite are eliminated in people through the use of Mectizan, the blackfly will have no source of them. The disease cycle between blackflies and people will end. River blindness will no longer be a threat to human health and well-being.

MERCK ESTABLISHED THE MECTIZAN DONATION PROGRAM

By and large, onchocerciasis patients cannot afford to pay for expensive drugs to treat their disease. Having an effective and safe drug accomplishes nothing if it is not used. Merck faced a dilemma.

Here is how Roy Vagelos, by now CEO of Merck, saw the situation[10]:

> There were only two possibilities. If we decided to sell Mectizan, it wouldn't reach those who needed it most regardless of how low we set the price. This was unacceptable for a company dedicated to improving human health. If, on the other hand, we decided to give it away, we would set a dangerous precedent for a pharmaceutical company that needed profits to sustain the sort of research and development that made Mectizan possible. Not only that, other groups suffering from virulent diseases such as malaria or AIDS might expect similar donations of medicines. Giving Mectizan away would be expensive for Merck in any case. At that time 18 million people were infected, and 80 million more were at risk worldwide. I worried that our giving the drug away might discourage some organizations from developing other drugs for impoverished countries. But I also felt that we were in a unique situation: we hadn't originally set out looking for a cure for river blindness, but we were now in a position to help millions of people.

Roy Vagelos did the right thing; he elected to donate the drug. "I decided Merck would give the drug free to any person endangered by river blindness anywhere in the world for as long as it was needed."

To get the drug to patients, Merck created the Mectizan Donation Program in 1987. Currently, more than 140 million doses of Mectizan are approved each year for onchocerciasis in Africa, Latin America, and Yemen. The world is better off for having ivermectin.

Ivermectin in Combination with Albendazole Is Effective Against Lymphatic Filariasis

Infection by the filarial roundworm *Wuchereria bancrofti* causes the human disease lymphatic filariasis, also known as *elephantiasis* (Figure 10.5). Some 120 million people are infected and, of them, 40 million are disfigured and incapacitated. The number of people infected dwarfs the population with onchocerciasis.

The *W. bancrofti* worms occupy the lymphatic system and disrupt normal immune function. The adult worms elicit the symptoms of the disease. Ivermectin kills only the microfilariae, disrupting reproduction of the worms but failing to relieve disease symptoms. The antiparasitic drug albendazole (a member of the same structural class as thiabendazole) kills the adult worm. Clinical trials have established that the combination of the two drugs is a safe and effective treatment for lymphatic filariasis.

In 1998, Merck expanded the Mectizan Donation Program to include lymphatic filariasis in many parts of the world, again at no cost to patients. Each year, 130 million Mectizan and albendazole treatments are approved for lymphatic filariasis in Africa and Yemen.

There is legitimate hope that, over time, both onchocerciasis and lymphatic filariasis can be eliminated completely from Earth. If so, the avermectins will have been key players in that triumph. For the present, they continue to contribute to human health in some of the poorest places on Earth at no cost to the victims.

11 Fludalanine
NICE TRY BUT NO HALLELUJAH

IT IS TIME that we experienced a tale of failure. As I have stated repeatedly, most drug discovery efforts fail. Choice of the wrong target dooms the effort from the start. Screening may fail to turn up actives, and molecular design may do no better. Given active molecules, medicinal chemistry efforts to improve properties may fail. Senior management's heavy hand may terminate a promising effort. If one gets as far as development, a safety or efficacy issue may derail the project. Then, too, competitors may outrun you or financial support may dry up. There are many ways to fail, not so many to succeed.

All five stories told so far have been successes: finasteride, ACE inhibitors, statins, imipenem/cilastatin, and the avermectins. Chapter 12 provides another example of a success: the gliptins. In this chapter, however, I pull together the threads of a failure: fludalanine. It is the most interesting failed drug discovery story that I know. There is much to learn from it, particularly about problem solving. It has a couple of surprises.[1]

By way of background, Merck had set its mind on finding an effective orally active antibiotic, driven in substantial part by the insistence of Max Tishler, Merck's determined head of research. Orally active antibiotics are attractive. A patient with a bacterial infection may be treated in the hospital with an oral or a parenteral agent, one given by injection or inhalation. An injectable antibiotic may be given intravenously, intramuscularly, or subcutaneously. However, when the patient has been released from the hospital and sent home, it is convenient to have an antibiotic that can be taken by mouth, a tablet or capsule, for continued action against the infection. Ideally, an antibiotic should be available in both a parenteral (intravenous, intramuscular, or subcutaneous) formulation and an orally active formulation. In this way, the patient can be maintained on the same antibiotic when returning home from the hospital.

Merck was having trouble meeting its objective. Fosfomycin, an antibiotic isolated from the fermentation broth of a soil microorganism by Merck in collaboration with a Spanish natural products company, showed useful activity but was plagued by rapid resistance development. It is now used principally for urinary tract infections. Moving forward, Merck in-licensed pivampicillin, which is a prodrug form of the antibiotic ampicillin and is converted to ampicillin in the body. Pivampicillin has the advantage of increased oral absorption compared with ampicillin. This drug candidate, too, failed to meet Merck's objectives, as a result of apparent liver toxicity in early human testing (although the finding was very probably spurious). It was eventually approved in several countries but is not an important product and is not available in the United States.

Having had two drug development failures did not dampen Merck's lust for an orally active antibiotic. Fludalanine was the next molecule that had the potential to fill the need. As the story develops, you may be surprised at the lengths to which Merck scientists went in efforts to save the molecule. Persistence is a success factor in drug discovery; bull-headedness is not. It is often difficult to know where the dividing line lies. Here is the story. You can decide for yourself.

Fludalanine Targets the Synthesis of the Bacterial Cell Wall

I provided the background to antibiotic discovery and development in the Primaxin story in chapter 9. There is no need to repeat it here except to remind you that novel antibiotics, and particularly novel *classes* of antibiotics, are difficult to come by. Antibiotic resistance is a major health issue worldwide. We continue to need novel antibiotics effective against bacteria that are or have become multidrug resistant.

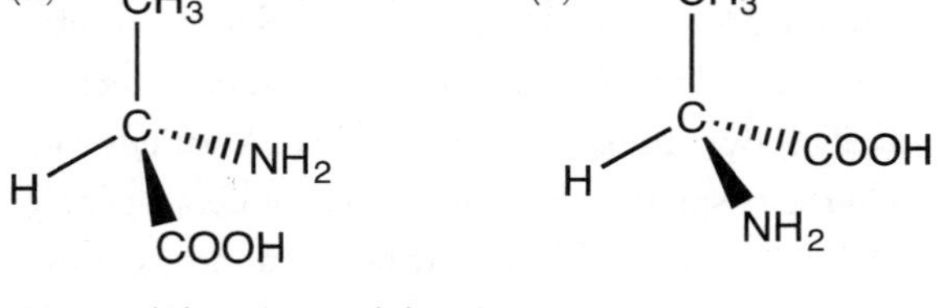

FIGURE 11.1 The Alanines. (A) D-alanine. (B) L-alanine.

The bacterial cell wall, because of its unique composition, is an attractive target for antibiotic action. Indeed, Fleming's seminal discovery of penicillin, and subsequent β-lactam antibiotics such as the cephalosporins (chapter 9) are directed against the last step in bacterial cell wall assembly. Fludalanine also exploits the synthesis of the bacterial cell wall as a target, but it does so a bit earlier in the process of cell wall synthesis.

The key understanding is that synthesis of the bacterial cell wall requires the amino acid D-alanine. Alanine is a very simple amino acid; only glycine has a simpler structure. Alanine (the three letter abbreviation is Ala or ala) is a common component of proteins. There is a subtle point here: alanine comes in two flavors, L and D. They are nonsuperimposable mirror images (Figure 11.1). Think about a pair of gloves. The gloves are composed of exactly the same building materials but are arranged differently in space. We might think of them as the D glove for the right hand and the L glove for the left hand. These gloves are nonsuperimposable mirror images, just as L- and D-alanine.

Proteins contain only the L-isomer, L-alanine. Human, animal, and bacterial proteins are composed only of L-amino acids; they contain no D-amino acids. In contrast, the bacterial cell wall peptidoglycan contains D-alanine and a second D-amino acid, D-glutamate. These exotic amino acids render the bacterial cell wall resistant to the action of certain mammalian enzymes that recognize L-amino acids only, which provides the bacteria protection against them.

D-alanine is sourced from abundant L-alanine along the pathway to cell wall synthesis. Specifically, the bacterial genome encodes two enzymes known as *alanine racemases* (AR1 and AR2; hereafter, simply AR) that catalyze the equilibration of L- and D-alanines (Figure 11.2).

The action of these enzymes, unique to bacteria, provides the only source the bacteria have for D-alanine. So, Merck scientists formulated a drug discovery hypothesis: an effective inhibitor of AR would be an effective antibiotic. Starving the bacteria for D-alanine should prevent bacterial cell wall synthesis and, therefore, bacterial replication. The absence of AR in the human enzyme repertoire is a plus in terms of safety. AR was adopted as the target of a target-based drug discovery project.

The next question was: how do we come up with an effective inhibitor of AR? At the least, Merck needed an inhibitor good enough to test the underlying hypothesis of the drug discovery effort.

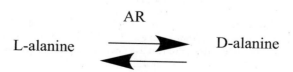

FIGURE 11.2 Alanine Racemases (AR) Catalyze the Equilibration of L- and D-alanines.

A Merck Fishing Expedition Lands an Interesting Catch

Motivated by the siren call of "new chemistry," Merck chemist Janos Kollonitsch had embarked on a systematic effort to add an atom or two of fluorine (F) to a series of biological intermediates in a search for novel molecules with useful biological activity. In this work, he used a technique known as *photofluorination*, which he pioneered. Kollonitsch is a Hungarian chemist who came to Merck after escaping from the Communist regime in Hungary.[2] His efforts to add fluorine atoms to small molecules using his new chemistry were not a pointless exercise. Many molecules useful in human medicine contain one or more fluorine atoms. Kollonitsch submitted his new fluorine-containing molecules for general biological testing.

Among the fluorinated molecules created in his laboratory was D-fluoroalanine, in which one atom of fluorine replaced an atom of hydrogen. Lou Barash was the chemist who first made D-fluoroalanine. Steve Marburg was the chemist who learned how to make useful quantities of it in the Kollonitsch laboratory.

D-fluoroalanine proved to be a potent inhibitor of AR. In fact, this molecule modifies AR chemically in a way that destroys its catalytic activity permanently. It is an irreversible inactivator of AR. Here is how this works.

D-fluoroalanine looks a great deal like D-alanine. Just one atom is different, and the fluorine atom, like the hydrogen atom, is quite small. In fact, it looks good to AR, and AR recognizes it as a substrate and acts on it. During the course of this process, a very cool thing happens. A highly reactive intermediate structure is formed (this intermediate cannot be formed from D-alanine itself). The reactive intermediate has two potential fates. First, it can just diffuse off the surface of the enzyme and end up in the enzyme environment. Second, it can react with the enzyme and kill it. That is, it forms a stable chemical bond with some atom of the enzyme and the resulting product has no catalytic

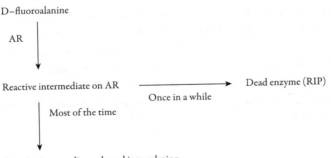

FIGURE 11.3 Mechanism of Irreversible Inhibition of Alanine Racemases (AR) by D-fluoroalanine. D-fluoroalanine forms an enzyme–substrate complex with AR, after which a chemically reactive intermediate is formed. Most of the time, this reactive intermediate dissociates from AR and ends up in the reaction solution, freeing AR to react again. Once in a while, however, the intermediate reacts chemically with the AR and kills it. We end up with a dead enzyme to which the reactive intermediate is irreversibly bound. Inhibition of this type is sometimes called *suicide inhibition* because the killer agent is formed by the action of the enzyme itself.

activity. Most of the time, the reactive intermediate diffuses off the enzyme, but from time to time it modifies AR and kills it. Have a look at Figure 11.3. Molecules that act in this way are known as *suicide enzyme inhibitors*. By acting on them as if they are the normal substrate, the enzyme forms a reactive intermediate that kills itself. Merck had its inhibitor to test the hypothesis: would D-fluoroalanine kill bacteria?

D-fluoroalanine Shows Antibiotic Potential

The initial evaluation of D-fluoroalanine for antibiotic activity proved negative. It was tested at a single high concentration. Not deterred, Kollonitsch sent a sample to biologist Fred Kahan for a second opinion. Here is another example of where the right assay yielded critical insights. Fred used a disc diffusion assay. A small disc of filter paper impregnated with the test material is applied to the surface of an agar plate covered with a layer of bacteria. As the test material diffuses away from the disc, it forms a concentration gradient. There is a high concentration of test material near the disc, but decreasing concentrations as the distance from the disc increases. If the test material is active in killing the bacteria, a clear zone remains around the disc. Farther out, where the concentration of test material is too dilute to be an effective antibiotic, bacterial colonies grow and cover the agar. This idea is illustrated in Figure 11.4. When Fred performed a disc diffusion assay with D-fluoroalanine, he obtained a remarkable result. Instead of the typical zone of inhibited growth extending out from the disc, the

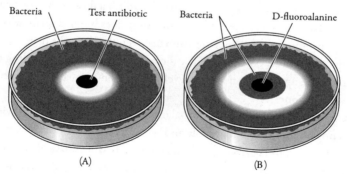

FIGURE 11.4 Two Examples of the Results of Disc Diffusion Assays. In both cases (A and B), the surface of an agar layer in a Petri dish is covered with a thin layer of bacteria. In the center of the agar is placed a small piece of filter paper impregnated with a test antibiotic. Over time, the test material diffuses away from the filter paper and into the agar. Concentrations of test antibiotic will be highest nearest the filter paper and diminish with increasing distance. (A) Typical result of a disc diffusion assay when the test antibiotic, perhaps penicillin, is active. A clear zone remains near the filter paper where bacterial growth has been prevented by high concentrations of the antibiotic. Farther out, the antibiotic is too dilute to be effective and bacterial colonies grow and cover the agar. (B) The result when the test substance is D-fluoroalanine. The result is remarkable. Near the filter paper, there is growth of bacterial colonies. A bit farther out, there is a clear zone where bacterial growth has been prevented by the antibiotic action of D-fluoroalanine. Still farther out, where the D-fluoroalanine is dilute, there is again growth of bacterial colonies. This is an illustration of self-reversal. The amazing point is the growth of bacterial colonies near the filter paper where the concentration of D-fluoroalanine is highest.

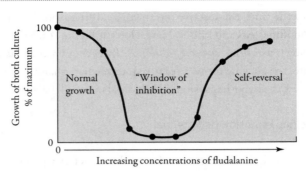

FIGURE 11.5 Dose–Response Profile for Bacterial Killing by D-fluoroalanine. At low concentrations of D-fluoroalanine, bacterial growth is inhibited, as expected. As the concentration is increased, bacterial growth is essentially stopped, the "window of inhibition." Remarkably, bacterial growth resumes at even higher concentrations of D-fluoroalanine, the phenomenon of self-reversal. This is the work of Fred Kahan and Helmut Kroop at Merck. (Source: Redrawn from Birnbaum J. 1982. In *Antibiotics in the Management of Infections: Outlook for the 1980s*, ed. A. G. Bearn New York: Raven Press.)

inhibited zone was offset from the disc by an inner ring of growth adjacent to the disc (Figure 11.4). Remarkably, D-fluoroalanine inhibited the growth of bacteria at intermediate concentrations but not at the highest concentrations near the disc. This phenomenon became known as *self-reversal*. There are two points to be made. First, D-fluoroalanine has antibiotic activity; second, the phenomenon of self-reversal makes it useless as an antibiotic.

There is a second way to observe the phenomenon of self-reversal. When taking the biological measure of a molecule, it is typical to establish a dose–response profile for activity (which is basically what the disc diffusion assay does but in a qualitative way). The dose–response profile defines the relationship between the dose of a drug and the consequent biological response. Normally, one expects an increasing dose of drug to elicit an increasing response up to some maximal level.[3] If this is not the case, then some serious thinking has to be done.

The dose–response profile for D-fluoroalanine as an antibacterial is provided in Figure 11.5. It is nothing short of remarkable and recapitulates the result seen with the disc diffusion assay shown in Figure 11.4. As expected, increasing doses of the drug initially increase bacterial killing. However, as the dose is increased further, the effect reverses and the bacteria begin to thrive again, just as we saw with the disc diffusion assay. At the highest doses tested, the bacteria grow almost as well as in the absence of the antibiotic. This phenomenon was observed with several strains of bacteria, both Gram positive and Gram negative. Merck scientists showed that at high concentrations of D-fluoroalanine, AR was completely inhibited, as expected. To the best of my knowledge, this observation has neither precedent nor progeny in the search for antibiotics. It cries out for an explanation.

Fred Kahan and Helmut Kropp provided the explanation. They proved that, at high concentrations of D-fluoroalanine, the enzyme involved in catalyzing the subsequent

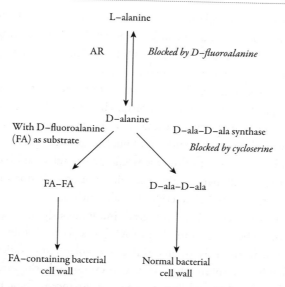

FIGURE 11.6 Biochemical Basis for the Phenomenon of Self-reversal by D-fluoroalanine. In the absence of D-fluoroalanine, alanine racemases (AR) catalyze the conversion of L-alanine to D-alanine, which is then converted to D-ala-D-ala by the enzyme D-ala-D-ala synthase. D-ala-D-ala then goes on to form a part of the structure of the normal bacterial cell wall. In the presence of D-fluoroalanine (FA), no D-alanine can be formed as a result of inhibition of AR. However, D-ala-D-ala synthase recognizes FA as a substrate and catalyzes the formation of FA-FA, which then goes on to form a part of the structure of an FA-containing bacterial cell wall. Only in the presence of both an inhibitor of AR, D-fluoroalanine, and an inhibitor of D-ala-D-ala synthase, cycloserine, is bacterial cell wall synthesis inhibited.

step in the bacterial cell wall pathway, D-ala-D-ala synthase, was fooled into accepting D-fluoroalanine as a building block in place of D-alanine. (Remember that ala is the three letter abbreviation for alanine.) The function of this enzyme is to link two D-alanine molecules together to make D-ala-D-ala, which is a cell wall building block used as the process of cell wall synthesis progresses. Instead of making D-ala-D-ala, the action of the enzyme creates fluoro-D-ala-fluoro-D-ala. The bacterial cell simply uses this building block in place of D-ala-D-ala and completes cell wall synthesis (Figure 11.6). The antibiotic becomes an integral part of a fluorinated cell wall (a Teflon-like cell wall!). Clearly, this would not do. What could be done about self-reversal?

Cycloserine Is a Potent Inhibitor of D-ala-D-ala Synthase

The enzyme D-ala-D-ala synthase is the problem. It is responsible for the first step in the incorporation of D-fluoroalanine into the bacterial cell wall (Figure 11.6). One possible way out of the self-reversal problem might be to use an inhibitor of that enzyme together with D-fluoroalanine. D-fluoroalanine would inhibit AR and an inhibitor of D-ala-D-ala synthase would prevent incorporation of D-fluoroalanine into the cell wall structure. The effect on cell wall synthesis should be synergistic, and self-reversal should be prevented. The effort to create a combination antibiotic containing two molecules rather than one is demanding. However, if D-fluoroalanine was going to survive as a product candidate, there seemed to be no other option.

This strategy had a precedent. The combination of trimethoprim and sulfamethoxazole is a successful product for treatment of certain bacterial infections. These antibiotics inhibit successive steps in the biosynthesis of a molecule known as *tetrahydrofolate*, which is an essential player in the synthesis of nucleic acids and proteins.

The corresponding situation in this case would be to find an antibiotic that inhibits D-ala-D-ala synthase and then combine it with D-fluoroalanine. The combination would inhibit successive steps in bacterial cell wall synthesis.

A molecule called *D-cycloserine*, discovered simultaneously in 1954 by Eli Lilly and Merck, was known to be a weak inhibitor of AR but a potent inhibitor of D-ala-D-ala synthase. In his disc diffusion assay, Fred Kahan was able to show quickly that the combination was effective. With an adjacent disc of cycloserine, the inner region of self-reversal was abolished and the inhibitory zone was enhanced. These results provided a very neat proof of concept.

Cycloserine has some safety issues, mainly in the CNS. It can cause headache, depression, psychosis, and seizures. These safety issues limit its clinical utility. They appear to derive from two sources. One is weak activity at specific receptors in the CNS; the other is instability in both solution and the solid state, giving rise to a toxic product. The latter issue is, in principle at least, resolvable by finding a way to stabilize cycloserine.

Merck chemist Norm Jensen discovered a stable derivative of cycloserine, named *pentizidone*, which liberates cycloserine slowly in the bloodstream and more rapidly in the kidney. Pentizidone is another example of a prodrug, a molecule without biological activity itself but which the body converts into one with useful activity. Merck selected pentizidone as a companion product for D-fluoroalanine, with the anticipation that it would be better tolerated than cycloserine itself. Whether this is, in fact, the case was not established because the clinical trials were of short duration before the compound's development was abandoned.

The expectation was that D-fluoroalanine and pentizidone would be synergistic in killing bacteria. The combination of the two enzyme inhibitors proved to have potent antibiotic activity in animal studies and abolished the self-reversal phenomenon, as expected. This combination was taken forward into development.

D-fluoroalanine/Pentizidone Has Attractive Antibiotic Properties

Subsequent investigations yielded encouraging results. The D-fluoroalanine/pentizidone combination proved highly effective as an antibiotic. Furthermore, it showed promise against a broad spectrum of bacteria, both Gram positive and Gram negative—a highly desirable property because, in most bacterial infections of humans, the causative organism is not known. So, it is usually best to have an antibiotic that kills lots of pathogenic species. Last, it is orally active; in fact, it is completely absorbed after oral administration. These were clear signs of promise, but there were two concerns beyond the issue of self-reversal.

Pentizidone also addressed a subtle problem with the combination of cycloserine and D-fluoroalanine. The elimination rate of cycloserine by the kidney is much slower than that of D-fluoroalanine. A potential consequence is that high concentrations of D-fluoroalanine would accumulate in the urine at times when the concentrations of urinary cycloserine remain low. This imbalance raises the potential for D-fluoroalanine self-reversal in the urine, with loss of activity against bacteria in the kidney. Fortunately, an early fraction of intact pentizidone is cleared rapidly by the kidney and dissociates quickly in the more acidic urine to provide cycloserine at levels that block self-reversal.

This situation provides a compelling example of the issues that must be faced in drug discovery and development after the desired primary activity—antibiotic activity, in this case—is established. These issues include pharmacokinetics—the profile of drug concentration in blood and key tissues or other bodily fluids as a function of time after dosing—drug metabolism, and safety. In this case, we need to get the concentrations of cycloserine and D-fluoroalanine matched in the urine and kidney over time in a way that avoids self-reversal and lack of efficacy for kidney infections. Pentizidone met that need; cycloserine did not. Next we encounter issues of metabolism and safety for the pentizidone/D-fluoroalanine combination.

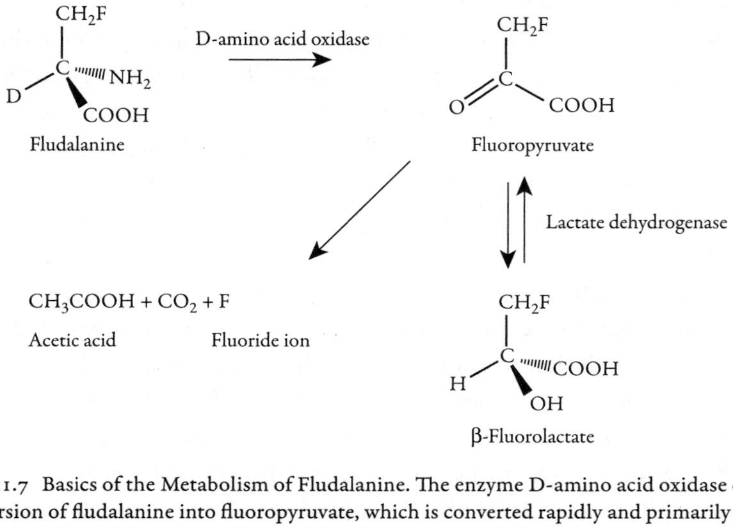

FIGURE 11.7 Basics of the Metabolism of Fludalanine. The enzyme D-amino acid oxidase catalyzes the conversion of fludalanine into fluoropyruvate, which is converted rapidly and primarily to 3-fluorolactate through the action of lactate dehydrogenase. It is 3-fluorolactate that elicits brain damage in long-term safety studies in experimental animals. The small pool of fluoropyruvate is converted, in part, to acetic acid, carbon dioxide, and fluoride ion. Although fluoride ion has toxic potential, it was not a problem in fludalanine safety assessment studies.

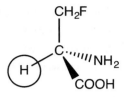

FIGURE 11.8 D-fluoroalanine. The circled hydrogen atom is lost during the metabolism of this molecule by D-amino acid oxidase. Replacement of this hydrogen atom by an atom of deuterium slows the rate of metabolism, ultimately slowing the rate of loss of fluoride in a subsequent step.

Although mammals do not use D-amino acids, they do have the ability to metabolize them through oxidation by the enzyme D-amino acid oxidase, found in liver, kidney, and white blood cells. In the case of D-fluoroalanine, metabolism by this enzyme results in the formation of fluoropyruvate (Figure 11.7). This molecule liberates its fluoride ion through subsequent metabolism. Fluoride presents a theoretical risk of tooth mottling in children.

Metabolic studies revealed that rodents treated with D-fluoroalanine did excrete small amounts of fluoride ion. Because metabolism of D-fluoroalanine carries with it some safety risk, it seemed desirable to limit it to the extent reasonably possible.

Fred Kahan came up with a novel way to do this. He recognized that D-fluoroalanine would be a target for the enzyme D-amino acid oxidase. During the course of acting on D-fluoroalanine, this enzyme would pull off the hydrogen atom circled in Figure 11.8. Although this reaction does not liberate a fluoride ion, the rate of the process determines the rate of release of fluoride ion from D-fluoroalanine in a later metabolic step. So, the question became: is there some way to slow down that process?

The answer is yes, and the way to do it is to replace the circled hydrogen atom (Figure 11.8) with a deuterium atom. We need a bit of chemistry here. Recall that hydrogen is the simplest of the elements. By definition, it contains just one proton in the atomic nucleus and one orbital electron. Hydrogen comes in three flavors, known as *hydrogen isotopes*. The most common form of hydrogen, H or ^{1}H, has no neutrons in the nucleus. Deuterium, D or ^{2}H, contains a nuclear neutron to accompany the proton. Tritium, T or ^{3}H, the third isotope of hydrogen, has two nuclear neutrons. Tritium is radioactive; deuterium is stable.

The basic difference between an atom of hydrogen and one of deuterium is mass. A neutron weighs just slightly more than a proton, so an atom of deuterium weights about twice as much as an atom of hydrogen (the electron weighs very little compared with a proton or neutron). For reasons based on mass difference, breaking a chemical bond between carbon and deuterium is slower than breaking a chemical bond between carbon and hydrogen. It follows that D-amino acid oxidase should act on deuterio-D-fluoroalanine (D) more slowly than on the protio (H) form. The technical name for the molecule containing the deuterium atom is *3-fluoro-2-deuterio-D-alanine*. It is the molecule known as *fludalanine*.

This expectation was borne out. In rodents dosed with fludalanine, the rate of liberation of fluoride ion from fludalanine was less than half that of D-fluoroalanine. On this basis, Merck decided to go with the fludalanine in clinical trials. I believe that this is the first example of an effort to influence the rate of metabolism of a drug by replacing a hydrogen atom with one of deuterium at a site of bond breakage. In principle, this might have saved fludalanine had release of fluoride ion been the critical safety issue. But it was not. Read on.

In short-term safety studies, the toxicity of fludalanine proved to be very low. The minimal toxicity that was seen reflected that of fluoride ion liberated from fludalanine by metabolism and was minimized by the incorporation of the deuterium atom into fludalanine. To provide the basis for multidose clinical trials, Merck moved to long-term safety studies. Another problem developed.

Long-Term Toxicity Studies Revealed a Serious Safety Problem

Multiple-dose studies over increasing periods of time in the clinic require long-term animal toxicity studies. Anticipating a clinical start, Merck began 3-month toxicity studies of fludalanine in rats and Rhesus monkeys. Necropsy revealed myelin vacuolation in the corpus callosum of the brain in both species. Histopathology resembled that of "spongy brain" found in infants and burn patients treated with hexachlorophene. Put in the most basic possible terms, there were tiny holes in the brains of the treated animals.

Some toxicity issues observed in animal studies are not a stop signal for the project. For example, a finding of a mild liver problem (hepatotoxicity) in animal studies might be accommodated by a series of liver function tests in clinical trial patients. What is ethically acceptable depends, of course, on the nature and severity of the observed toxicity, and the dose and the duration of drug treatment required to elicit the issue compared with the anticipated human dose and treatment duration.

The brain toxicity seen with fludalanine was another matter. Unlike routine liver function or kidney function tests, there is no brain function test that is likely to reveal toxicity in human studies until it has become irreversible. In the absence of a compelling explanation for the toxicity and a reasonable expectation that it would not be a problem in people, there was no way to move forward. Fludalanine had leaped the fluoride ion issue and had conquered the self-reversal issue, but seemed doomed by brain toxicity. However, this is not the end of the story.

A Metabolite of Fludalanine Causes the Brain Toxicity in Experimental Animals

The experimental finding that fludalanine elicited brain damage in animals did not necessarily mean that it was a toxic molecule. Most drugs are subject to metabolism in the bodies of animals and humans. In other words, they are acted on by enzymes and converted into other molecules, termed *metabolites*. These molecules are derivatives of the drug that are ultimately excreted in the urine or feces. As noted earlier, D-amino acid oxidase metabolizes fludalanine, converting it into other products. One or more of the metabolites, rather than the parent compound, could have been responsible for the toxicity. The basic aspects of the metabolism of fludalanine are presented back in Figure 11.7.

So it was with fludalanine. In an effort led by Fred Kahan, Merck scientists determined the course of metabolism of fludalanine. An important metabolite is known as *3-fluorolactate*. It was demonstrated in toxicity studies that 3-fluorolactate, not fludalanine, caused vacuolation in the brain of rats and Rhesus monkeys. Merck had identified the culprit, which opened the door to a potential rescue of the fludalanine/pentizidone combination. Opening this door, however, depended—in an important way—on the clinical use of the combination as an antibiotic. These drugs are usually taken for short, defined periods of time, perhaps a week to 10 days. Most bacterial infections are treated successfully during this time frame, the patient is cured, and drug administration is stopped. Had the combination been intended for chronic use, such as blood pressure control, I believe that Merck would have simply given up on the project without worrying about which molecule caused the toxicity. Chronic use would have posed too great a risk to patients to be acceptable ethically.

Here Is the Process Adopted to Establish Threshold and to Negotiate a Patient Safety Factor

CNS toxicity with drugs is common and frequently acceptable when it is thought to be reversible. However, histopathological changes in the brain such as those seen with 3-fluorolactate are unlikely to be reversed on stopping the drug. In cases such as this one, the principal is that we must remain well below the minimum toxic level of the causative agent—3-fluorolactate.

The issue that required resolution was whether levels of 3-fluorolactate in people taking fludalanine at effective antibiotic doses pose any risk of irreversible brain damage. The way to find the answer was to correlate the blood levels of 3-fluorolactate with brain damage in experimental animals (known as *toxicokinetics*) with the blood levels of 3-fluorolactate in patients on effective doses of fludalanine/pentizidone, which was done.

The first step was to identify the threshold blood level of 3-fluorolactate associated with brain damage in experimental animals, which Merck scientists determined. It was agreed that further development of fludalanine/pentizidone was contingent on maintaining 3-fluorolactate blood levels in patients at no more than 10 percent of the threshold level, a 10-fold safety factor. That is, the goal was to keep blood levels of 3-fluorolactate in people to one-tenth or less of the threshold level in experimental animals. With this agreement in place, clinical trials were an acceptable way forward with the combination.

Phase 1 clinical studies with fludalanine/pentizidone were started in *normal volunteers*. Doses were increased gradually until they reached the level expected to be effective in treating infections. At this dose, blood levels of 3-fluorolactate were below the maximum acceptable level. The FDA granted permission to proceed to phase 2 trials in *patients*.

However, the first phase 2 trial in patients with bronchitis and chronic obstructive lung disease ended the life of the fludalanine/pentizidone combination. Some patients with bronchitis revealed double the expected blood levels of 3-fluorolactate, which was unacceptable. In retrospect, this observation might have been expected. Lactate itself is metabolized by oxidation. Patients with bronchitis and chronic obstructive lung disease have lung function problems (hypoxia) and are compromised in their ability to carry out oxidative metabolism. They have higher than normal levels of blood lactate. Because D-fluorolactate metabolism is also oxidative, it is not surprising that blood levels in these patients were higher than those for volunteers with normal lung function. In 1984, Roy Vagelos, president of Merck Research, announced that the development of fludalanine/pentizidone had been stopped. The long path forward had reached a dead end.

Note that the deuterium atom in fludalanine might have rescued the molecule by slowing the formation of 3-fluorolactate sufficiently. The effect was in the right direction (by about a factor of two), but was not sufficiently profound. Had the deuterium effect been large enough, fludalanine/pentizidone might have been an important antibiotic.

Fludalanine had cleared the fluoride release hurdle, the self-reversal hurdle, and the brain toxicity issue in healthy people, but could not clear the hurdle in the population for which it was intended—people with bacterial infections.

Work Continued

Although the fludalanine/pentizidone combination failed as an antibiotic for reasons of patient safety, work with it had established that AR in bacteria was an excellent target for antibiotic activity. Art Patchett and his chemists went back to the inhibitor design drawing board and came up with other designs based on enzyme chemistry.

Basic Research Provides Insights for Enzyme Inhibitor Design

There are a lot of biochemists in the academic world involved in basic research. The results of this research provide insights that may be exploited to advance human health. One focal point of academic biomedical research has been, and is, to understand in detail how enzymes achieve their amazing catalytic efficiency. This is how I spent the bulk of my life in the academic world. It has gotten to the point where we understand enzymes in reasonably good detail.

Half a century ago, Esmond Snell, a professor of chemistry at the University of California at Berkeley carried out a series of investigations on how enzymes that depend on a derivative of vitamin B_6 for activity perform their chemistry. These enzymes include many racemases—specifically, AR. Insights derived from Snell's work permitted Merck scientists to understand the mechanism of inhibition of AR by D-fluoroalanine. Inspired by the fludalanine story, Art Patchett (the same guy who is responsible for Merck's ACE inhibitors and who found lovastatin in a fermentation broth at Merck's Rahway, New Jersey, site) later rationally designed a small family of effective inhibitors of AR based on Snell's mechanistic insights. Given these insights, Merck might well have discovered fludalanine through a process of rational drug discovery.

The best of the Patchett designs, known as D-chlorovinylglycine, is a far more effective inhibitor of AR than fludalanine and cannot exhibit the self-reversal phenomenon, nor would one expect it to be incorporated into the bacterial cell wall in place of D-alanine. However, it is a failure as a potent antibiotic, presumably because it cannot penetrate the bacterial cell wall to gain access to AR in the cell wall synthetic process.

It is easier to make great enzyme inhibitors than great drugs.

12 Diabetes Breakthroughs
JANUVIA AND JANUMET

I DO NOT recommend this, but let's just suppose that for breakfast you had a waffle drenched in maple syrup, a large glass of orange juice, and coffee with two teaspoons of sugar. What happens next?

Your breakfast is full of carbohydrates. The waffle contains complex carbohydrates (starches) from the flour, and the orange juice, syrup, and sugared coffee contain simple carbohydrates (sugars). The sugar in orange juice and syrup is mostly fructose, and that in your coffee is sucrose. In the intestines, the starches are slowly broken down into the sugar glucose; the sucrose is split into equal amounts of glucose and fructose. The fructose can be converted to glucose in the liver. The sugars from the orange juice, syrup, and coffee enter the blood quickly; those from the starches in the waffle enter more slowly. However, they all act to increase blood glucose levels.

A basic principle of human physiology is summed up in one word—homeostasis, which simply means that when the normal metabolic status of the human body is changed in some way, the body responds by restoring normality. When something changes, the body fights back to eliminate or minimize the change. This is what happens when blood glucose[1] is elevated in response to a meal. Here is how.

Insulin Acts to Restore Blood Glucose Levels to a Normal Range

The pancreas is a medium-size organ in the abdomen that secretes enzymes into the gut to aid in digestion, and endocrine hormones into the bloodstream to control some aspects of metabolism. The pancreas responds to sugar entering the bloodstream by secreting the peptide hormone insulin (see Figure 3.2) into the circulation. Insulin is made in specialized cells of the pancreas known as the *beta cells* of the islets of Langerhans. Insulin has a critical role in the regulation of blood glucose levels.[2] Acting through its receptor (see Figure 4.6), insulin causes glucose in the blood to be taken up by muscle and fat cells, reducing the blood glucose level. Liver cells also take up glucose but that process is not (entirely) insulin dependent. In muscle and liver, insulin also promotes the conversion of glucose into glycogen, a starchlike storage substance for carbohydrates. These events are summarized in Figure 12.1.

There is a question here: how does the pancreas know that a load of glucose has been dumped into the bloodstream? The glucose sensor is a somewhat curious enzyme known as *glucokinase*, one member of a small family of enzymes that catalyze the conversion of glucose to glucose-6-phosphate, consuming a molecule of ATP—the energy source that drives the reaction—in the process. Glucokinase is largely localized in the liver, where it acts to balance glucose metabolism between energy storage and energy production, depending on the needs of the organism. However, there is glucokinase in the beta cells of the pancreas. When glucose levels rise, glucokinase responds by increasing the levels of glucose-6-phosphate, which, in turn, initiates a series of reactions leading to insulin secretion. The details of this process are pretty well understood but a bit wide of the mark for our story. Suffice it to say that glucokinase is the glucose sensor that regulates insulin secretion.

There is a second question: how does insulin regulate the uptake of glucose by muscle and fat cells? The key protein here is the glucose transporter type 4 (GLUT4).

When GLUT4 is localized on the cell membrane of muscle or fat cells, glucose is transported through the cell membrane into the cell interior where it is trapped by conversion

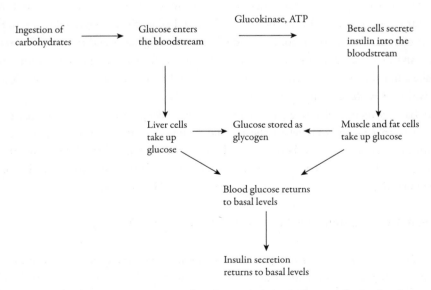

FIGURE 12.1 Metabolic Events That Normalize the Level of Blood Glucose After Intake of a Carbohydrate Meal. ATP, adenosine triphosphate.

to glucose-6-phosphate by a different enzyme in the glucokinase family—hexokinase. In the absence of insulin, about 95 percent of GLUT4 is stored inside vesicles within the cell. A limited amount of glucose can enter cells through the small number of GLUT4 molecules exposed on the cell surface. Insulin acts to markedly increase the fraction of GLUT4 molecules available for glucose transport. It follows that the rate of flow of glucose into the muscle or fat cells will increase in the presence of insulin. Once the level of insulin decreases as normal levels of blood glucose are restored, the GLUT4 molecules are once again stored within vesicles inside the cell and the rate of glucose transport into muscle and fat cells returns to its basal level. Have a look at Figure 12.2 for a summary of these events.

As I noted above, the transport of glucose into liver cells is independent of insulin. In these cells, the glucose transport protein is GLUT2, rather than GLUT4. The extent of localization of GLUT2 on the surface of liver cells is independent of insulin.

In summary, a dietary load of glucose entering the bloodstream is sensed by glucokinase in the beta cells of the pancreas. That initiates a process resulting in increased secretion of insulin into the blood. That, in turn, increases the number of GLUT4 transporters on the surface of muscle and fat cells. Increased uptake of glucose in these cells results, accompanied by insulin-independent glucose uptake by liver cells. The end result is that blood glucose levels return to their normal level, as does the blood level of insulin.

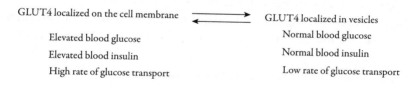

GLUT4 localized on the cell membrane   GLUT4 localized in vesicles

    Elevated blood glucose   Normal blood glucose

    Elevated blood insulin   Normal blood insulin

    High rate of glucose transport   Low rate of glucose transport

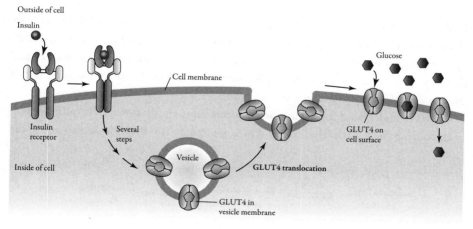

FIGURE 12.2 Insulin-Dependent Movement of GLUT4 between the Cell Membrane and Intracellular Vesicles. Under conditions of normal blood glucose levels, blood insulin levels are also normal and most of the glucose transporter GLUT4 is localized within vesicles within the cellular interior. The rate of glucose transport from the bloodstream into the cells will be low. When blood glucose levels are elevated, blood insulin levels are also elevated, in response to the glucose load, and GLUT4 is relocated to the cell membrane. As a result, the rate of glucose transport from the bloodstream into the cells is increased, tending to normalize blood glucose levels.

In muscle and liver, storage of glucose in the form of glycogen is enhanced. That is the way things are supposed to work following your carbohydrate-laden breakfast.

Glucagon Plays a Role in Blood Glucose Control

Insulin is not the whole story for regulation of blood glucose levels. The α (alpha) cells of the pancreas secrete a second peptide hormone—glucagon. Glucagon is a single chain of 29 amino acids. Its actions are largely opposed to those of insulin.

Let's suppose that you have digested your big breakfast thoroughly and have gone for a long, vigorous bicycle ride in hilly country. The energy demands of your bike ride will decrease blood glucose levels as glucose is metabolized to generate ATP, the energy currency of the cell. This situation will not do. We need some way to restore normal levels of glucose in this glucose-depleted blood.

Glucose inhibits the secretion of glucagon from the pancreas. As glucose levels decrease, this inhibition is relaxed and glucagon is secreted. Acting through the glucagon receptor, it initiates a process that results in the breakdown of glycogen to glucose in the liver, restoring blood glucose to its normal level. Glucagon is a counterregulatory hormone to insulin, as illustrated in Figure 12.3. Between the actions of the two, blood glucose levels are restored to normal, whether too high or too low.[3]

Unhappily, not everything works as it should for everyone. When it does not, pathology in the form of some disease may result. In our case, that disease is diabetes mellitus.

Diabetes Is a Failure of Normal Blood Glucose Control

Diabetes comes in two forms, unimaginatively termed *type 1* and *type 2 diabetes*. Both are characterized by too much glucose in the blood, which is called *hyperglycemia*. Type 1 diabetes is an autoimmune disease in which the human immune system attacks and destroys the beta cells of the pancreas. Because these cells are the only human source of insulin, untreated type 1 diabetes is a fatal disease.

Efforts to understand and treat type 1 diabetes led to the discovery of insulin by Frederick Banting and Charles Best in Toronto in 1921/1922. This breakthrough is one of the

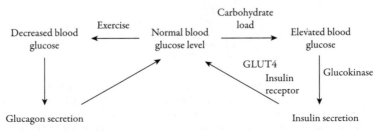

FIGURE 12.3 The Opposing Effects of Insulin and Glucagon on the Regulation of Blood Glucose Levels. It is critical to have agents that act to reduce and increase blood glucose levels, because an uncorrected excursion in either direction is associated with pathology.

great stories in human medical history.[4] In 1982, Michael Bliss published an intimately detailed history of the discovery of insulin, updated in 2007: *The Discovery of Insulin: 25th Anniversary Edition*. The opening words of this book follow:

The discovery of insulin at the University of Toronto in 1921–22 was one of the most dramatic events in the history of the treatment of disease. Insulin's impact was so sensational because of the incredible effect it had on diabetes patients. Those who watched the first starved, sometimes comatose, diabetics receive insulin and return to life saw one of the genuine miracles of modern medicine. They were present at the closest approach to the resurrection of the body that our secular society can achieve, and at the discovery of what has become the elixir of life for millions of human beings around the world.[5]

Insulin is a really important molecule in human physiology! About 5 to 10 percent of patients with diabetes have type 1. The great majority has type 2, a more subtle disease.

HYPERGLYCEMIA DEFINES TYPE 2 DIABETES

Type 2 diabetes is typically diagnosed in middle age, although it is being seen increasingly in younger individuals. There are three diagnostic criteria for type 2 diabetes in asymptomatic persons, pretty much agreed on by the American Diabetes Association and WHO. The first is a fasting blood glucose level equal to or greater than 126 mg/dL. This is the glucose level at which problems with the retina (retinopathy) that can lead to blindness—a complication of diabetes—may occur. The second is a blood glucose level equal to or greater than 200 mg/dL 2 hours after an oral loading dose of 75 g of glucose (a glucose tolerance test used in the diagnosis of diabetes). The last criterion is a finding that 6.5 percent of hemoglobin or more is glycated at a specific location on the hemoglobin molecule. Glycated hemoglobin is usually referred to as *HbA1c*, and is the result of a chemical reaction between groups on hemoglobin and glucose in the blood. Fasting blood glucose and a glucose tolerance test give data on blood glucose levels at a point in time. In contrast, levels of HbA1c provide a measure of glucose levels averaged over a 2- to 3-month time period. There is nothing magical about these limits. They simply reflect the views of experts in the field of diabetes based on a wealth of experience with patients with diabetes and the complications of their disease.

TYPE 2 DIABETES HAS A LONG ASYMPTOMATIC STAGE

Type 2 diabetes is preceded by a long period of increasing inability to control blood glucose levels properly. This condition is termed *prediabetes*. Individuals with prediabetes are usually asymptomatic, although about 5 percent show symptoms of microvascular disease (discussed later). They are, however, at increased risk of developing diabetes over time, and of developing the symptoms of that disease.

The American Diabetes Association defines prediabetes as (1) a fasting blood glucose level in the range of 100 to 125 mg/dL or (2) blood glucose level after a glucose tolerance test in the range of 140 to 199 mg/dL or (3) an HbA1c level in the range of 5.7 to 6.4 percent.

Individuals with prediabetes have a 5 to 10 percent annual risk of developing diabetes. This risk is 5 to 10 times greater than it is for individuals with normal fasting glucose levels or normal glucose tolerance. Diagnosis of prediabetes is important because it has been shown in controlled clinical trials that the risk of developing diabetes can be reduced by lifestyle modifications, including better diet, weight reduction, and an increase in exercise level.

Type 2 Diabetes Is a Huge Medical Problem

Type 2 diabetes is an epidemic. In the United States, there are an estimated 26 million people with the disease, more than 8 percent of the population, and another 79 million with prediabetes. Worldwide, diabetes affects 240 million people. Estimates suggest that this figure will grow to 380 million by 2025, a huge increase.

Diabetes is closely associated with obesity.[6] As the fraction of humankind that is obese increases as a result of a calorie-rich, nutrient-poor diet coupled with a sedentary lifestyle, so does the incidence of diabetes. The sequelae of diabetes include heart disease, stroke, kidney failure, blindness or impaired vision (retinopathy), and pain or a tingling sensation (neuropathy). After heart disease and cancer, diabetes is the number three killer of people in the United States. The annual cost of diabetes in the United States runs into the hundreds of billions of dollars.

There are three hallmarks of diabetes: insulin resistance, beta-cell dysfunction, and overproduction of glucose by the liver. All contribute to increasing the levels of glucose in the blood. Insulin resistance is a condition in which the body produces insulin but its effect on muscle, fat, and liver cells is compromised. The pancreas responds by secreting more insulin to try to cope with elevated blood glucose levels. Many patients with diabetes have above-normal levels of circulating insulin; the insulin just does not work well. Eventually the beta cells exhaust their ability to meet the demand for insulin— beta-cell dysfunction—and diabetes results. The overproduction of glucose by the liver is a failure to respond appropriately to insulin. Insulin becomes less able to shunt glucose to storage as glycogen, which results in release of glucose into the circulation.

From the standpoint of treatment, the question is: how do we restore the normal, healthy controls of blood glucose to their nondiabetic state? Scientists have been working on this problem for many years and we have a number of drugs with which physicians treat diabetes. However, none are entirely satisfactory and further progress remains to be made. The story of Januvia and Janumet provided in this chapter is one important response to this need.

We Have Several Families of Drugs to Treat Type 2 Diabetes

The first option for patients with type 2 diabetes is to attempt to control their blood glucose levels through diet, weight control, and exercise. If a patient can achieve this end, there is no need to consider drug therapy. Unhappily, most patients with type 2 diabetes cannot control their blood glucose levels through behavioral modification, and drug therapy is indicated. Clinical trials have demonstrated that improved blood glucose control results in lowered risks for the *microvascular* complications of diabetes: nerve damage (neuropathy), kidney disease (nephropathy), and damage to the retina (retinopathy). The better the control, the lower the risk of experiencing these consequences of diabetes. The data that link improved glucose control to a lower risk of the *macrovascular* complications of diabetes—coronary heart disease and stroke—are less compelling. Patients with diabetes are at increased risk for cardiovascular disease.

There is no drug that cures type 2 diabetes. The objective of drug therapy is to normalize blood glucose levels and prevent the complications of diabetes. Ideally, we want to normalize blood glucose levels in a glucose-dependent manner just as normal physiology does. Remember, secretion of insulin by the pancreas is in response to a glucose load.

For type 1 diabetes, insulin is life-saving. It also finds use in type 2 diabetes, but usually after oral agents (discussed later) have failed to control blood glucose levels properly. Insulin is given by injection[7] and there are a variety of formulations available. The principal downside of giving insulin is that the administration is not glucose dependent. That is, the insulin does not respond to a glucose load and is not controlled by a glucose load as it is in normal physiology. In an attempt to mimic normal physiology, insulin injections are correlated with food intake and physical activity. The most serious consequence of insulin use is hypoglycemia, a blood glucose level less than 70 mg/dL. The symptoms of hypoglycemia include rapid heartbeat, nervousness, headache, hunger, trembling, sweating, confusion, and weakness. If not treated by restoration of normal blood glucose levels, hypoglycemia can be fatal. The threat of hypoglycemia is not limited to insulin, as noted later.

Metformin (Glucophage) is first-line therapy and the most widely used oral drug for type 2 diabetes. It is a member of the biguanide class of drugs for diabetes, known as a class as *antihyperglycemics*. Its main effect is to reduce glucose output by the liver. It also reduces insulin resistance, thus increasing the uptake of glucose by muscle cells and decreasing insulin required to normalize blood glucose levels. Metformin is widely used in combination with other diabetes drugs including insulin. Metformin does not cause weight gain. Metformin may elicit problems in the gastrointestinal tract, such as diarrhea, nausea, upset stomach, and vomiting. Metformin's position as first-line therapy for type 2 diabetes has direct consequences for the structure of the clinical trials for sitagliptin. As detailed later, sitagliptin was used in combination with metformin in some clinical trials.

The most popular of the sulfonylurea class of antihyperglycemics include glibenclamide (aka glyburide; marketed as Diabeta and Micronase), glipizide (Glucotrol), and glimepiride (Amaryl). Agents in this class work by blocking an ATP-sensitive potassium channel in beta cells of the pancreas, resulting in increased secretion of insulin. There is a risk of hypoglycemia with sulfonylureas because the insulin secretion is not glucose dependent. Sulfonylureas also have the potential to cause weight gain.

Finally, there is the glitazone class of antihyperglycemics (thiazolidinediones).[8] The class includes pioglitazone (Actos) and rosiglitazone (Avandia). Concerns about safety, including bladder cancer, have resulted in a marked decrease in the use of these drugs. Avandia has been withdrawn from the market in Europe and has restricted labeling in the United States as a result of cardiovascular safety concerns. A more nearly complete collection of oral antidiabetics before the discovery of the gliptins is provided in Table 12.1.

There are two points here. The first is that there are several families of drugs that are taken orally that aid in the control of blood glucose levels by different mechanisms in patients with type 2 diabetes. The second is that these drugs leave something to be desired, both in terms of efficacy and safety. There was a clear medical need for improved drugs for the control of blood glucose levels in type 2 diabetics.

Scientists interested in working on this problem faced the universal question: what should we do? As usual, the answer was not clear, but there were intriguing suggestions accompanied by worries, which brings us to the sitagliptin story.

Glucagonlike Peptide 1 Has Provocative Properties

Glucagonlike peptide 1 (GLP-1) is a peptide composed of a chain of 37 amino acids cleaved from a larger product of the proglucagon gene. An enzyme activates it by chopping off the first six amino acids of the chain. It is formally known as GLP-1(7–37). We are

TABLE 12.1

Collection of Oral Drugs for the Control of Type 2 Diabetes Before the Discovery of the Gliptins

Chemical class	Members	Molecular target	Principal mode of action
Biguanides	Metformin	Unknown	Decrease glucose production by the liver
Sulfonylureas	Glibenclamide, glipizide, glimepiride, glicalzide	ATP-dependent potassium channel in beta cells	Increase insulin secretion by beta cells
Thiazolidinediones	Pioglitazone, rosiglitazone	PPARγ	Increase peripheral sensitivity to insulin

ATP, adenosine triphosphate.

going to refer to it simply as *active GLP-1*. GLP-1 is secreted from certain intestinal cells (L cells) in response to carbohydrates, proteins, or fat in the gut—that is, it is secreted in response to a meal.

Active GLP-1 has two highly interesting properties in the context of a search for a treatment for type 2 diabetes. First, it induces the secretion of insulin from the pancreas *in a glucose-dependent manner*. Molecules in this class are known as *incretins*. GLP-1 is the most important incretin. Second, it represses the secretion of glucagon, an antagonist of insulin, from the pancreas in a *glucose-dependent manner*. The effects on glucose-dependent secretion of insulin and glucagon result in a potent lowering of blood glucose levels *when blood glucose is elevated*. On the surface of things, GLP-1 looks like a great drug for diabetes.

However, there is a problem—if not, there would be no story here. The problem is that active GLP-1 is degraded quickly to biologically inactive products in the circulation. Specifically, about half of all the active GLP-1 in the blood is degraded every 2 minutes. In other words, active GLP-1 is said to have a half-life in the blood of 2 minutes. So, in 2 minutes, half of it is gone; in 4 minutes, half of the remainder is gone; and in 20 minutes, more than 99 percent is gone. The half-life of GLP-1 is much too short to be consistent with using it as a drug; it would have to be delivered into the bloodstream by continuous infusion to be effective. At the same time, GLP-1 is a great research tool.

ACTIVE GLP-1 NORMALIZES ELEVATED BLOOD GLUCOSE LEVELS

Although GLP-1 is useless as a drug for diabetes, it is a splendid tool for testing the underlying hypothesis of a drug discovery project: a stable molecule with the biological properties of active GLP-1 would be a useful drug for controlling blood glucose levels in patients with type 2 diabetes.

This hypothesis was put to the test in 1997. In a short-term study, GLP-1 was given to patients with type 2 diabetes by continuous infusion into the bloodstream. Blood glucose levels were normalized in both the fasting state and after a meal. Blood glucose normalization was an exciting result and encouraged scientists to take the underlying hypothesis seriously. A subsequent study completed in 2002 confirmed and amplified this result. A 6-week continuous infusion of GLP-1 in patients with diabetes was shown to decrease fasting blood glucose and HbA1c levels significantly. This result was important, because it showed that the beneficial effects of GLP-1 in diabetes could be sustained over time. Scientists followed up on these observations.

There are two ways to think about what you might do to exploit this opportunity. One possibility is to find a molecule with the biological properties of active GLP-1 with a long half-life in blood. The second option is to identify the enzyme responsible for the degradation of GLP-1 in the blood and discover an inhibitor for that enzyme. Such an inhibitor would extend the lifetime of active GLP-1 and its beneficial effects. Scientists have traveled both avenues. Let's have a look at them, starting with the GLP-1 agonists.

GLP-1 AGONISTS ARE USEFUL AGENTS FOR THE CONTROL OF BLOOD GLUCOSE IN PATIENTS WITH TYPE 2 DIABETES

The search for stable GLP-1 agonists—molecules with the same biological properties as active GLP-1 itself[9]—has resulted in the introduction of new antidiabetic drugs into clinical practice (Figure 12.4).

The first GLP-1 agonist to be approved by the FDA for treatment of type 2 diabetes was exenatide, marketed as Byetta. Exenatide is a modified version of a peptide isolated from the saliva of the Gila monster—natural product chemistry proves its worth again! Peptides are either degraded in the gut or do not cross the gut wall and enter the general circulation. So, they cannot be administered effectively orally. Exenatide is given twice a day about an hour before meals by subcutaneous injection. As expected, exenatide helps to control elevated blood glucose levels and reduces the level of HbA1c, and its use is accompanied by significant weight loss, which are all positive effects. Exenatide can be used in combination with other medicines for type 2 diabetes including metformin. On the less positive side is the inconvenience of twice-daily subcutaneous injections and some side effects, the most significant of which are nausea and vomiting.

Exenatide is available in a formulation that releases the active agent over extended time periods—Bydureon. Bydureon needs to be injected subcutaneously only once a week, a considerable convenience for patients compared with Byetta.

The second GLP-1 agonist approved for control of type 2 diabetes is liraglutide, marketed as Victoza. Liraglutide is a chemically modified form of GLP-1 with a blood half-life of 13 hours. It is injected subcutaneously once a day before a meal.

Both Bydureon and Victoza have been found to cause thyroid tumors in rodents. It is not known whether they have the potential to cause such tumors in people. Both products carry a black box cancer warning in their labeling. Many physicians hesitate to prescribe drugs with black box warnings if an alternative considered to be safer is available.

In sum, the strategy of creating GLP-1 agonists has proved successful with new products approved for treatment of diabetes. None of these are recommended for first-line treatment of the disease but they are useful additions to the family of medicines useful for the patient with type 2 diabetes.

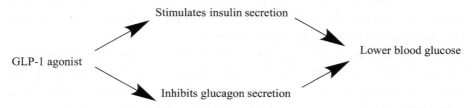

FIGURE 12.4 Actions of a GLP-1 Agonist. Stimulation of insulin secretion coupled with inhibition of glucagon secretion results in lower blood glucose, the desired outcome for patients with type 2 diabetes mellitus.

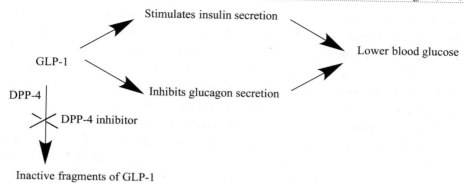

FIGURE 12.5 Action of a DPP-4 Inhibitor. By protecting GLP-1 from degradation to inactive fragments, an inhibitor of DPP-4 would prolong the beneficial actions of GLP-1.

Now we need to get on with exploring the second strategy: the discovery of inhibitors for the enzyme responsible for the rapid degradation of GLP-1 in the blood, which will get us to the sitagliptin story.

DIPEPTIDYL PEPTIDASE 4 DEGRADES ACTIVE GLP-1

During the 1990s, it became clear that an enzyme named *dipeptidyl peptidase 4 (DPP-4)* catalyzes the degradation of active GLP-1 in the bloodstream. It was not clear at that time that DPP-4 is the only enzyme to carry out this function. Nonetheless, it was chosen as the molecular target for inhibitor discovery. The underlying hypothesis of the project is direct: a specific inhibitor of DPP-4 would prove safe and effective for treatment of type 2 diabetes. Such an inhibitor would protect active GLP-1 from degradation by DPP-4, prolonging its beneficial actions (Figure 12.5).

In 1999, Merck initiated a drug discovery project based on this hypothesis. Nancy Thornberry led the project biology and Ann Weber did the same for chemistry. As always, the discovery of sitagliptin was the result of work by many scientists. Here is their story.

The Choice of DPP-4 as the Molecular Target Came with Concerns

As I emphasized earlier, the choice of molecular target in a drug discovery project is critical. Many proteins are not "druggable." That is to say, having chosen a "nondruggable" protein target, you are doomed to failure at the outset. The sooner you know you are on this path, the better. There are lots of interesting and useful things to do; continuing down a path to nowhere is not one of them. Scientists involved in drug discovery attempt to test the underlying hypothesis of their work as early and often as reasonably possible.

DPP-4 is not obviously a druggable target. Here are the concerns. First, DPP-4 was believed to have multiple functions in human physiology, which is a worry from a safety standpoint. Ideally, one would like to have a molecular target that does just one thing. This quality makes it a great deal easier to predict the outcome of altering its action. Modulation of the action of targets that have many functions is likely to result in multiple pharmacological changes, some of which may prove to be adverse effects. Second, in the test tube, DPP-4 was known to act on multiple substrates, not just GLP-1. These included immunoregulatory, endocrine, and neurological peptides. It was entirely possible that inhibiting the action of DPP-4 on GLP-1 would create a favorable outcome for patients with diabetes, but that doing the same for some other substrate would have adverse consequences—another safety concern. Last, there was reason to believe that DPP-4 was involved in stimulating T cells, key components of the immune system. A DPP-4 inhibitor would, then, have the potential to compromise immune function.

It follows that the choice of DPP-4 as molecular target for the diabetes program was not without risk, but then the only way to know that your target of choice is sound is to wait until someone else proves the point. When that happens, lots of pharmaceutical companies join the race to try to come in second or third. It is more fun, and generally more profitable, to be the pioneer.

The risk associated with this project was reduced by the observation in 2000 that mice lacking DPP-4 developed normally and were healthy.[10] Merck scientists subsequently confirmed this observation in their own laboratories.

Merck was not the first pharmaceutical company to focus on creating inhibitors of DPP-4. In fact, when Merck got its project under way, there were already two companies with DPP-4 inhibitors in clinical trials: Novartis, a large multinational company, and Probiodrug, a small, privately held German company.

Merck Elected to In-License Two DPP-4 Inhibitors from Probiodrug

Given the competitive environment—DPP-4 inhibitors as well as GLP-1 agonists—in the clinic and moving forward, Merck elected to in-license two DPP-4 inhibitors from Probiodrug in 2000. The in-licensing occurred at a time when Merck had an active medicinal chemistry effort under way on DPP-4 inhibitors as well as a screening effort of the Merck compound collection for the same objective.

This is by no means unusual and the nature of this deal is typical: large pharmaceutical houses have money and small ones have attractive molecules, so money may be exchanged for molecules. The costs of clinical trials required to gain approval to market a drug are frequently enormous—hundreds of millions of dollars in some cases—although the cost depends on the therapeutic category. Getting to market is usually not an option for a small company; they simply do not have and cannot raise the necessary money. So, they are frequently willing to exchange molecules for upfront cash and additional financial rewards as the molecules proceed along the path to registration. These deals can be structured in any number of ways, but this is not the topic at hand. Suffice it to say that Merck and Probiodrug came to a mutually satisfactory set of terms.

The Probiodrug molecules were attractive. Single-dose clinical studies showed that the lead molecule was well tolerated, increased the blood level of active GLP-1, and reduced the excursion of blood glucose from normal levels after a meal in normal volunteers. In a small number of patients with type 2 diabetes, Probiodrug had shown that their lead molecule enhanced insulin secretion and improved glucose tolerance in single-dose studies. These are the measures one would look for in an effective DPP-4 inhibitor. Merck in-licensed the lead Probiodrug molecule and a close relative. I am going to call them PBD-1 and PBD-2.[11] The two are equally potent inhibitors of DPP-4, have similar half-lives in the blood, and are equally effective in a glucose tolerance test in obese mice. This turns out to be important, as discussed next.

Despite their attractions, the two molecules from Probiodrug failed in safety assessment studies at Merck: upon 5–6 weeks of treatment with PBD-1 in dogs, mortality and profound toxicities occurred at doses > 25 mg/kg/day. These toxicities included anemia, thrombocytopenia, splenomegaly, and multiple organ pathology affecting the lymphoid system and gastrointestinal tract. Put more simply, some of the dogs died. Others lived but with below-normal hemoglobin, below normal platelets, enlargement of the spleen, and lots of problems in the immune system and the gut. Those were the findings for PBD-1.

How about PBD-2? In a similar set of safety assessment studies, PBD-2 proved to be at least 10 times *more toxic* than PBD-1. Although this was bad news for PBD-2, it was good news in a sense for the Merck drug discovery program—a successful failure. As noted earlier, the two PBD molecules were equally effective as DPP-4 inhibitors and equally effective at affecting the consequent biology. The fact that one is 10 times more toxic than the other suggests that inhibition of DPP-4 is *not* the underlying cause of the toxicity. If it was, PBD-1 and PBD-2 should have been equally toxic, contrary to experimental observation. These results were surely encouraging, but left open the question of what caused the toxicity. Of course, development of PBD-1 and PBD-2 was ended.

Sometimes, events coalesce in a highly productive way; the stars are aligned properly. So it was with resolution of the safety issue associated with PDB-1 and PBD-2. Just as Merck scientists were dealing with this issue in 2000, others reported the discovery of two new enzymes coded for in the human genome—DPP-8 and DPP-9—which was one useful outcome, out of a great many, of the Human Genome Project.

Thus, DPP-4 is a member of a small family of related enzymes that includes DPP-8 and DPP-9. There are two others as well—quiescent cell protein dipeptidase (QPP) and fibroblast activation protein (or FAP)—although they play lesser roles in our story. It seemed entirely possible that PBD-1 and PBD-2 might inhibit DPP-8 and/or DPP-9 as well as DPP-4, and that such inhibition might elicit toxicity. That is a testable idea and Merck scientists set about testing it.

It turns out that PBD-1 and PBD-2 inhibit DPP-8, DPP-9, and QPP (as well as DPP-4). They are equally potent inhibitors of QPP, but PBD-2—the more toxic inhibitor—proved to be about 10 times more potent against DPP-8 and DPP-9 than PBD-1. This result accords very neatly with the observed toxicities and strongly suggests that the toxicity is associated with inhibition of DPP-8 and DPP-9. Merck scientists tested this suggestion rigorously. They discovered molecules that inhibited DPP-4 only and demonstrated the molecules to be free of the safety liabilities of PBD-1 and PBD-2. They also discovered molecules that inhibited DPP-8 and DPP-9 only, and showed that these shared the safety issues of PBD-1 and PBD-2. Case closed. These results were great, because they indicated to Merck scientists what they needed to do: find a potent inhibitor of DPP-4 that does not inhibit either DPP-8 or DPP-9. Finding such an inhibitor became the objective of the drug discovery effort.

As noted earlier, Novartis and Probiodrug had a head start—5 years in the case of Novartis—on Merck in the race to discover useful inhibitors of DPP-4. During these years, Novartis had no productive way to deal with safety issues related to inhibition of DPP-8 and DPP-9 because these enzymes had not yet been discovered. Timing can be everything in drug discovery and Merck happened to get it just right.

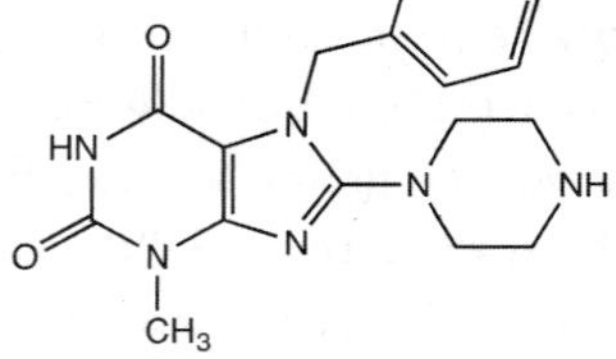

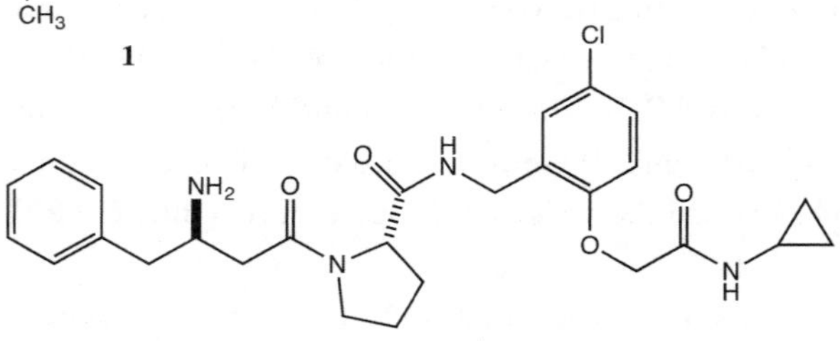

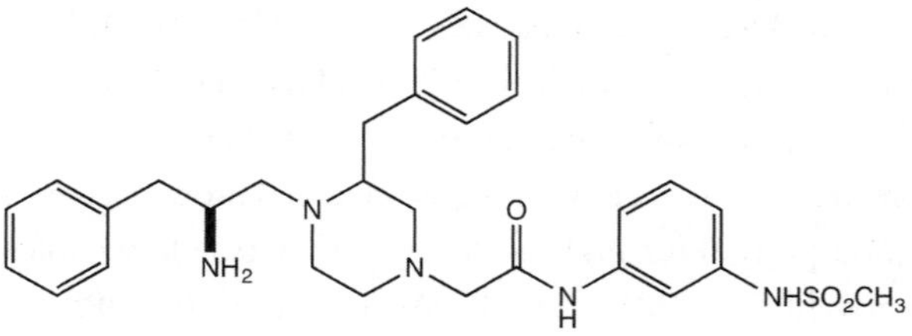

FIGURE 12.6 Three Active Inhibitors of DPP-4 Discovered When Screening the Merck Compound Collection.

Medicinal Chemistry Efforts Yield a Lead Compound

As noted earlier, compound screening and medicinal chemistry efforts at Merck were begun before in-licensing the Probiodrug molecules, and this work had continued. During the process, a potential lead compound had been discovered but abandoned for unacceptable levels of inhibition of DPP-8 and DPP-9 after that liability had been uncovered through the work with the Probiodrug compounds.

In the same structural class, further work yielded potent inhibitors of DPP-4 that greatly reduced the potential to inhibit DPP-8 and DPP-9. However, these inhibitors showed poor bioavailability (less than 1 percent) in rats.[12] Further work in this class failed to improve the oral bioavailability and the class was set aside.

While these chemistry efforts were ongoing, screening of the Merck compound collection was completed. Three compounds were identified that merited further work; they are labeled **1**, **2**, and **3** in Figure 12.6.

As a rule, I have tried to avoid writing chemical structures in these stories. I'm including the structures in Figure 12. 6 because I simply do not know how to tell the sitagliptin story without them. Even if you lack a chemistry background, I believe that there are matters here that are understandable and worth knowing. Hang in.

Compound **1** was obtained commercially and, therefore, was available to all pharma houses. Merck chemists made several novel molecules related to compound **1** with no breakthrough in potency. In part as a result of a concern that others might have discovered this active, Merck curtailed further work on this molecule. This concern proved to be well founded; the marketed DPP-4 inhibitor linagliptin (Figure 12.7) is in the same structural class as compound **1**.

Note that the boxed side of linagliptin (Figure 12.7B) is closely related to the structure of compound **1** (Figure 12.7A). A second marketed DPP-4 inhibitor—alogliptin—is also structurally related to compound **1**, although the relationship is less obvious.

Hits **2** and **3** provided the basic structural information that Merck chemists needed to get to sitagliptin. Let's see how that worked.

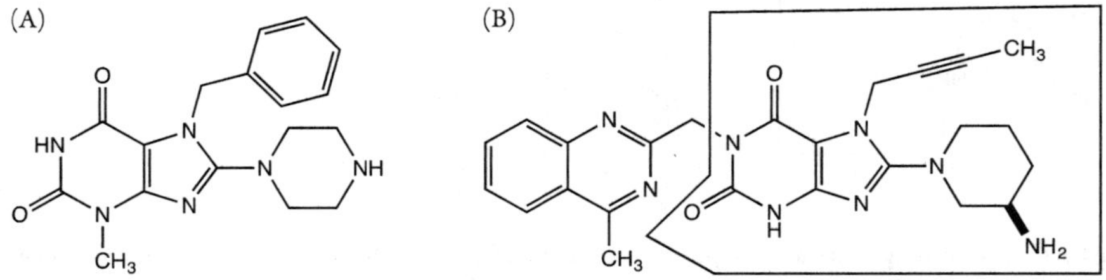

FIGURE 12.7 (A) Compound **1**. (B) Linagliptin. Note the structural similarity between compound **1** and the boxed atoms in linagliptin.

Near the beginning in a drug discovery process, when molecules are at an early stage of optimization for the desired properties, some parts of hits are more important than others. So an initial goal is to figure out which of the substructures within a molecule are key for activity. Doing this is a process of exploration; selected parts of a hit are deleted or modified in some way and the resulting molecules are tested—for example, for potency as inhibitors of DPP-4. With the benefit of medicinal chemistry experience and a good bit of back-and-forth between chemists and biologists, you can pinpoint the key structures within a molecule that are critical for activity. Merck chemists explored compounds **2** and **3** in this way and were able to identify a key structural unit within each hit. They then did the next logical thing and pulled these structural units together into one molecule, which is not trivial. There are usually several ways to pull different structural units together and, worse, none of them may work. However, this effort did succeed (Figure 12.8).

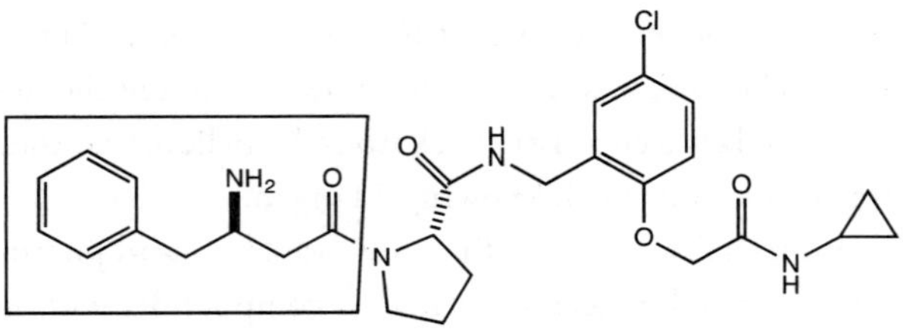

2

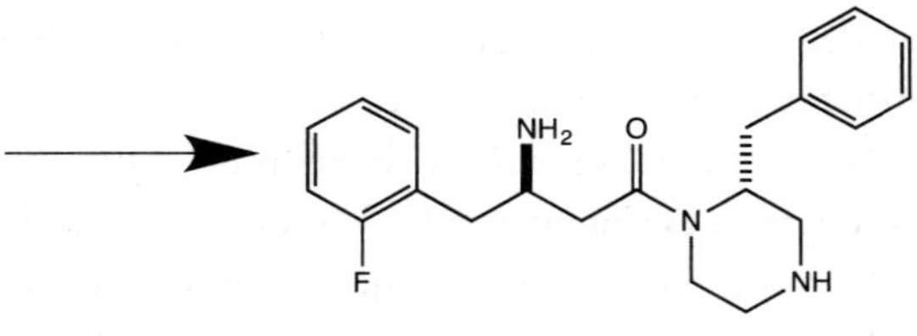

4

A lead molecule with attractive properties

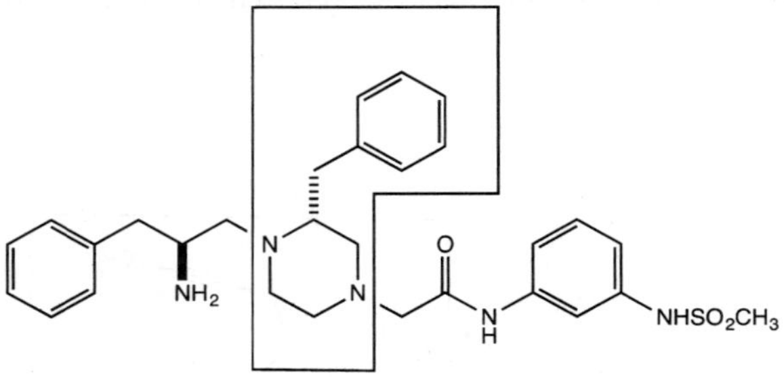

3

FIGURE 12.8 Joining Critical Groups of Atoms from Hits into a Novel Molecule Generates a Lead. Note that compound **4**, a lead molecule, is created by pulling together the boxed atoms in compounds **2** and **3** plus a fluorine atom. Compound **4** is far superior as a DPP-4 inhibitor to either compound **2** or compound **3**.

The key structural units in compounds **2** and **3** are enclosed in boxes in Figure 12.8. Now link together the boxed structures in compounds **2** and **3**, add a fluorine atom, F, and you get compound **4**.

Compound **4** is a lead molecule, derived from hits **2** and **3**. Compound **4** was about 15 times as potent as compound **2** and 80 times as potent as compound **3** as an inhibitor of DPP-4—big steps in the right direction. Beyond that, compound **4** was specific for DPP-4, having minimal inhibitory activity against DPP-8 or DPP-9, and was highly water soluble—another attractive property. Last, compound **4** was a molecule of modest size, smaller than either compounds **2** or **3**, which is hugely important because it provides chemists with the opportunity to add bulk in the form of new atoms or groups of atoms without getting beyond the usual size boundary for orally active molecules. In short, compound **4** had a family of pharmacological properties that qualified it as a lead molecule.

There were two related problems with compound **4**. First, it was highly susceptible to metabolism by liver enzymes. That is, in the presence of liver enzymes, compound **4** is converted to other molecules. Metabolism is a problem for two reasons: it shortens the lifetime of compound **4** in the circulation, and one or more of the metabolites may cause safety problems. Merck scientists were able to identify the six-member ring containing two nitrogen atoms (a piperazine ring) as the site of metabolism. Second, it had poor oral bioavailability. So, the goal of further chemical exploration was to modify compound **4** in a way that kept all the good properties—potency, specificity, water solubility—while building in good oral bioavailability and stability to liver enzymes.

Merck chemists solved the liver enzyme metabolism problem by fusing a five-member ring to the site of metabolism, which stabilized it and generated a new lead molecule, compound **5**, which retained potency, specificity, and water solubility, and gained metabolic stability (Figure 12.9). Oral bioavailability remained a problem.

Figure 12.10 provides the chemical highway from compound **5** to sitagliptin. The key chemical transformation is from compound **5** to compound **6**, in which a CH_2CH_3 group is replaced by a CF_3 group. They key here is that compound **6** had good oral bioavailability while maintaining all the good qualities of compound **5**.

Previous work at Merck had identified ways to improve the potency of inhibitors of DPP-4 by rearranging the fluorine atoms on the phenyl ring on the left-hand side of the molecule. This knowledge was applied to compound **6** to yield compound **7**. This change improved potency several fold and gave an additional boost to oral bioavailability. Thus, biological profile of compound **7** was taken.

This work established that compound **7** was potent, highly specific for inhibition of DPP-4, devoid of activity in a screen of 150 receptors and enzymes, had good to excellent oral bioavailability in five species, had an acceptable half-life, was effective in an oral glucose tolerance test in lean mice, and was clean in exploratory safety assessment studies in rats and dogs. It was approved as a preclinical development candidate.

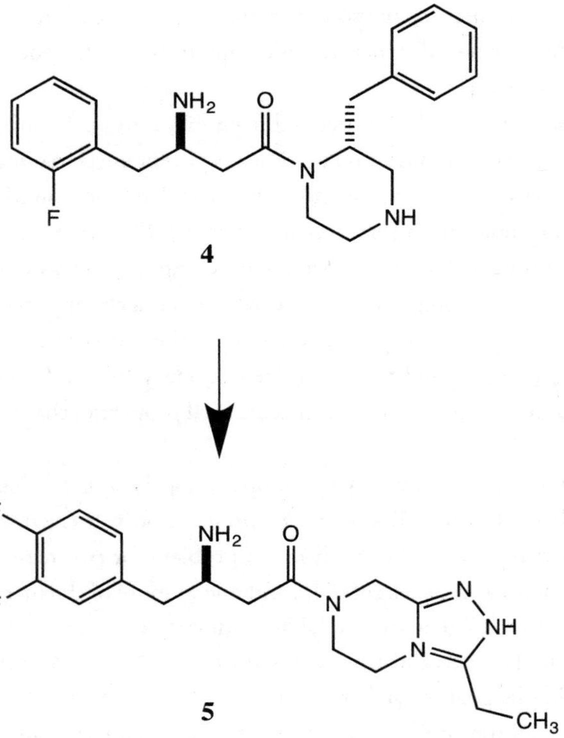

FIGURE 12.9 Key Structural Modifications to Compound 4 Yielded a Molecule with Improved Metabolic Stability: Compound 5. Note the rearrangement of fluorine atoms on the ring to the left and the addition of a new ring structure on the right side of compound 4.

But work continued. A simple derivative of compound 7 proved to have slightly improved potency compared with compound 7 and, more important, was devoid of the modest effects on heart rate and blood pressure in dogs seen with high doses of compound 7, which was a significant improvement in safety profile. The new compound was *sitagliptin*, which was also approved for development and prioritized ahead of compound 7. Figure 12.11 provides a look at sitagliptin complexed with DPP-4.

A great many new molecules were made in the search for sitagliptin, which is typical of drug discovery efforts. Chemists may make a couple thousand new molecules before finding one suitable for clinical trials (or failing to find one suitable for clinical trials). Beyond that, 90 percent of all molecules that enter clinical trials eventually fail. However, these discarded molecules do not disappear beyond the event horizon of some black hole of molecules. They become part of the growing collection of Merck molecules and will be screened in future drug discovery efforts. Molecules of no use for one application may find use for another. After all, the two key hits, 2 and 3, that provided structural elements for sitagliptin were molecules made in other drug discovery projects that proved critical to the success of this one.

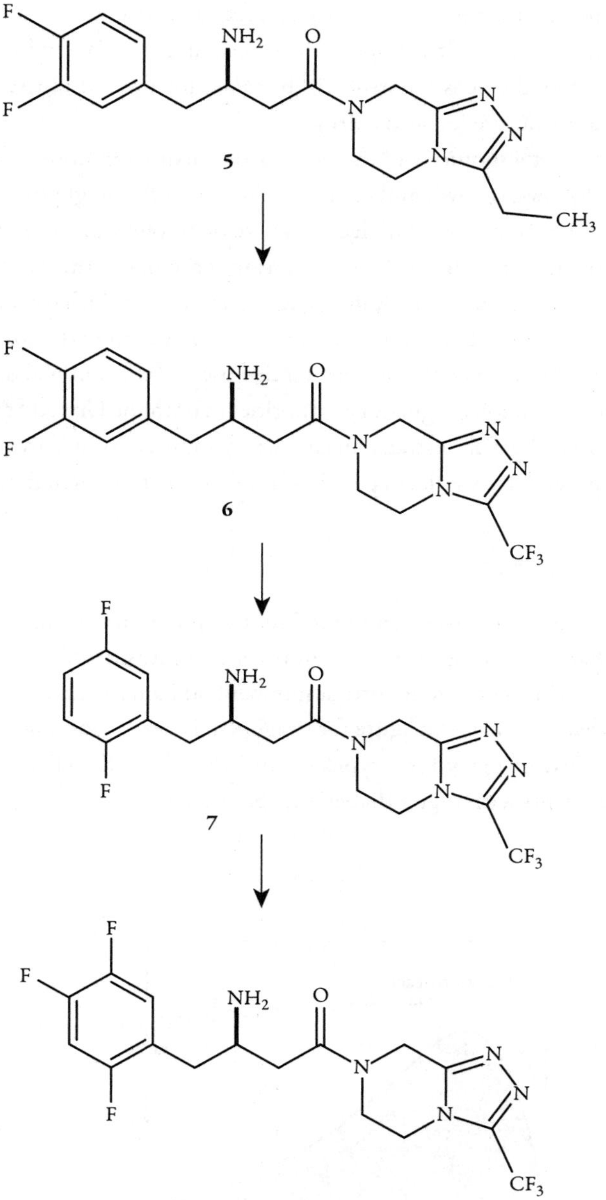

FIGURE 12.10 Key Chemical Advances Leading from the Lead Molecule **5** to Sitagliptin.

Here Is the Preclinical Development Program for Sitagliptin

As mentioned, everything that must be done between selecting a molecule for development and gaining approval for a clinical trial is termed *preclinical development*. I have not dealt with this area of drug discovery and development in much detail in the previous tales of drug discovery. Now is the time to do so.

The most important aspect of this work is safety assessment, necessarily carried out with experimental animals. It is important to recognize that a drug to assist patients with diabetes in blood glucose control will be taken for the rest of their life. This fact creates a high safety hurdle for such a drug.

To begin, I present a summary of the safety assessment work done in support of sitagliptin. In what follows, you will find mention of toxicities for sitagliptin, and they should come as no surprise. Basically, all molecules are toxic in one way or another given high enough doses for enough time. The safety question is the size of the window between effective doses in animal models for the intended use and the minimal toxic dose. The larger that window is, the better. In clinical trials, the question becomes the size of the window between effective doses in people for the intended use and the minimal toxic dose. Quite beyond these considerations, regulatory authorities in both the United States and Europe demand that toxicity be demonstrated in laboratory animals. Animal toxicology provides a window into potential toxicities in people—a clue of what to watch for in clinical trials.

SINGLE-DOSE TOXICITY

The first goal of the safety assessment program was to identify the minimal single dose of sitagliptin that would induce lethality in rodents: a measure of acute toxicity. In my personal opinion, this is not an important number and I am a bit surprised that it is still measured. Be that as it may, the figure for sitagliptin is near 2,000 mg/kg, which translates to a toxic dose of 140 g for a typical adult! Note that the effective daily dose of sitagliptin for patients with type 2 diabetes is about 0.1 g.

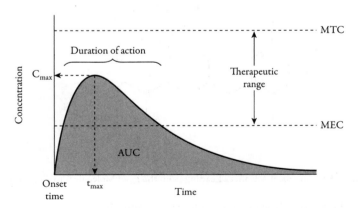

FIGURE 12.12 Typical Plot of the Concentration of Drug in Blood as a Function of Time After a Single Oral Dose. The area under the curve is a measure of the exposure to the drug. The onset time is defined as that time when the blood concentration reaches the minimal effective concentration (MEC). The duration of action of the drug is the time between the onset time and the time when the blood concentration falls below the MEC. The therapeutic window is the difference between the MEC and the maximal tolerated concentration (MTC).

REPEAT-DOSE TOXICITY

Observations in repeat-dose toxicity studies are far more important that those of single-dose studies. Blood drug levels are measured during these studies, so we get two measures of exposure: the dose of the drug and the associated blood levels. In Figure 12.12, I provide a typical plot of blood drug level against time for an oral dose of some drug. The blood level first increases as a function of time as the dose is absorbed and finds it way into the bloodstream. Eventually, the maximum concentration of drug in the blood is reached then it declines as the drug is eliminated. The area under curve is the key measure of drug exposure. In clinical trials, the blood levels of drug in people are measured and the drug exposure is measured as the area under the curve and is compared with that in animals that elicited toxic effects. Now let's look at a summary of the repeat-dose toxicity studies in support of sitagliptin.

Repeat-dose studies of sitagliptin were carried out in mice, rats, and dogs for periods ranging between 2 weeks and 53 weeks. Minimal signs of toxicity were observed, enough to satisfy regulatory authorities and clean enough to ensure that sitagliptin merited clinical trials in people. Later, a 14-week safety study (as opposed to a study designed to reveal toxicity) was carried out in monkeys at drug exposures approximately 20-fold higher than the drug exposure at an effective dose in people without evidence of toxicity. Overall, sitagliptin had a remarkably clean safety profile in these studies, and this profile carried over to patients with type 2 diabetes.

GENOTOXICITY

Genotoxicity studies are designed to detect gene mutations, direct damage to DNA, or chromosomal aberrations. Sitagliptin was clean in all these studies.

CARCINOGENICITY

Carcinogenicity studies were carried out for the lifetime of mice (about 21 months) and rats (about 24 months) at the highest doses of sitagliptin the animals could tolerate. All tissues were then examined for tumors at the conclusion of the study. Sitagliptin was clean in mice but did elicit liver tumors in rats at a drug exposure 60-fold greater than that for human use, apparently a result of sitagliptin-induced inflammation at the high doses used. Sitagliptin does not pose any carcinogenic risk to patients with type 2 diabetes at the recommended dose.

REPRODUCTIVE AND DEVELOPMENTAL TOXICITY

Reproductive and developmental toxicity studies are undertaken to look for the potential to cause birth defects—teratogenicity. No signs of reproductive toxicity were observed in rats or rabbits. Sitagliptin has not been studied in pregnant women and it is not recommended for use in women during pregnancy unless a clear need can be demonstrated. Sitagliptin has the potential to appear in the milk of lactating women, and caution should be exercised in this setting.

LOCAL TOLERANCE

In local tolerance studies the effort is to detect evidence of skin irritation or skin sensitization to irritants. Sitagliptin was clean in these studies.

The central point here is to recognize the lengths to which pharmaceutical companies must go to ensure that a compound is acceptable in safety terms for a clinical trial. The safety profile for sitagliptin is excellent, with no significant issues. Sitagliptin jumped the preclinical safety profile hurdle with ease.

HERE IS THE PHARMACOKINETIC PROFILE FOR SITAGLIPTIN

The issues considered during pharmacokinetic studies are extent of oral bioavailability, half-life in the blood, distribution within the body, metabolism to other products, and route of excretion. In rats and dogs, sitagliptin is absorbed rapidly and bioavailability is high after oral administration. Sitagliptin distributes widely throughout the body after oral administration, crosses the placenta in rats and rabbits, and appears in the milk of lactating rats. Metabolism of sitagliptin is minimal in rats, dogs, and rabbits. The metabolites that are formed are inactive as DPP-4 inhibitors. Sitagliptin was excreted largely unchanged in urine and feces. So here, too, we had a remarkably attractive set of properties of sitagliptin. Everything looked good.

FORMULATION

As noted back in chapter 5, the pills and tablets that patients take contain more than drug. Other ingredients are chosen from a list of compounds known to be safe for human consumption. There are multiple possible reasons for including them in the formulation, such as stability to light, heat, and moisture; optimal oral bioavailability; quick onset of action; compatibility with high-speed tableting machines; and mechanical strength of the tablet. For example, a tablet of sitagliptin contains microcrystalline cellulose, dibasic calcium phosphate, croscarmellose sodium, magnesium stearate, and sodium stearyl fumarate in addition to the active drug. It is important that the final formulation be used in the definitive phase 3 clinical trials.

SCALE-UP

Scientific activities before preclinical development generally require modest amounts of the drug molecule. When development work starts, this is no longer true. Instead of working on a scale of grams, one works on a scale of kilograms—1,000-fold greater.

A department sometimes known as *Process Development* takes over the synthesis of the drug material, which almost invariably means finding a new synthetic route to the product. Little attention needs to be paid to issues of the cost of synthesis at the gram scale, but this is no longer the case on the kilogram scale. A cost-effective synthetic route needs to be devised. Beyond that, there are some chemical procedures that work at the gram scale that simply cannot be scaled up 1,000-fold or more. These steps need to be replaced.

The cost of drug substance is not the only issue. Environmental concerns enter the picture. All chemical reactions produce waste products in addition to the desired product. Proper disposal of these waste products can be expensive and damaging to the environment, and a threat to human health. In an effort to deal with these issues, a branch of chemistry known as *green chemistry* has developed during the past couple of decades and flourished. The central point is to find chemical strategies that limit the production of undesired products.

In the United States, the Environmental Protection Agency presents Presidential Green Chemistry Challenge Awards annually to recognize new technologies that minimize pollution by reducing or eliminating hazardous waste in industrial production. In 2010, Merck and Codexis shared such an award for innovation in the synthesis of sitagliptin. Specifically, Merck and Codexis collaborated in the design of a custom enzyme that allowed for commercial-scale production of sitagliptin. Use of this custom enzyme increased the yield of sitagliptin while reducing the generation of waste products.

After a drug is approved for marketing, the synthetic route may be reconsidered again. Now we are not talking about kilogram scale; we are talking about perhaps tons of drug substance. However, the synthetic route cannot be altered without informing and getting the approval of the FDA in the United States and corresponding regulatory agencies abroad.

Here Is the Clinical Development Program for Sitagliptin

I am going to spend more time developing the story of the clinical program for sitagliptin than I have in our other tales. The sitagliptin program is much more recent than the others and more nearly reflects the environment in which clinical studies are now performed (although new guidance from the FDA for development of diabetes agents has been received since the licensing of sitagliptin). Beyond that, the clinical studies in support of sitagliptin are a model of focus and efficiency. There are lessons to be learned here and it seems worthwhile to take a bit of time to understand them.

There were a host of clinical studies undertaken for sitagliptin, and describing all of them makes no sense, so I will focus on key examples. They fall into six categories.

1. *ADME studies*: ADME studies are *a*bsorption, *d*istribution, *m*etabolism, and *e*xcretion studies. They answer the following questions: what fraction of a given

dose reaches the systemic circulation? How is the drug distributed among the tissues and organs of the body? Into what metabolites, if any, is the drug converted in the body? What is the route of excretion—urine or feces—for the drug and its metabolites? These studies provide basic information about how the drug is handled in the body, but they do not bear directly on either efficacy or safety (although safety is always monitored in clinical trials). ADME studies are usually considered to be part of phase 1 clinical trials.

2. *Pharmacokinetic studies*: Pharmacokinetic studies define the profile of the concentration of the drug in the blood as a function of time and dose, both after increasing single and multiple doses. Pharmacokinetic studies generally involve the testing of doses *greater* than those used in subsequent phase 2 and phase 3 studies, providing important safety information. A key quantity is the terminal half-life of the drug in blood because this determines, in part, the frequency of dosing. Longer half-lives translate to less frequent dosing. These studies are also part of phase 1 clinical trials.

3. *Clinical studies supporting the initial NDA*: These studies include the ADME and pharmacokinetic studies and the full range of safety and efficacy trials— phase 1 through phase 3—required to support the indications in the initial labeling of the drug (see chapter 5).

4. *Clinical studies supporting subsequent NDAs*: Additional clinical trials are often conducted after the drug is on the market after acquiring approval for the initial indications in its original formulation. These trials provide the data needed to gain other indications. For example, the initial NDA for ACE inhibitors (see chapter 7) gained approval for treatment of high blood pressure. Subsequent NDAs supported their use for congestive heart failure and other indications. These studies are also used to support new formulations, new strengths, and additional routes of administration, and to provide additional data on use in specific populations, such as the elderly.

5. *Clinical studies exploring uses that are likely to occur in clinical practice*: Physicians are not limited to prescribing medicines for approved uses or for treatment regimens explored in getting to an NDA. Hence, clinical studies are often undertaken to examine the consequences of uses not defined in the product labeling but that are likely to occur in medical practice.

6. *Everything else.*

The outcome of ADME clinical trials mimicked findings in experimental animals. Specifically, sitagliptin was well absorbed after oral administration, with about 87 percent of the dose appearing in the systemic circulation. Absorption was rapid, with peak blood concentrations reached in 1 to 4 hours. Food had no effect on the absorption of sitagliptin, so it can be taken with food or while fasting—a convenience for patients. Second, sitagliptin had a rather long terminal half-life in blood, near 12 hours, suggesting once-a-day or twice-a-day dosing. Steady-state drug levels were reached within 3 days of dosing and the drug did not accumulate in the blood. Third, metabolism of sitagliptin was limited. Most of the drug was excreted unchanged in the urine—that is, by way of the kidneys. The limited metabolism was also a plus in that the body is not exposed to metabolites that may carry a safety risk. Last, sitagliptin was widely distributed in the human body after oral administration. Everything here was just as one might have hoped. Things are usually not this clearly favorable.

The Key Concept in Sitagliptin Phase 1 Studies Was Target Engagement

Test tube studies proved that sitagliptin is a potent and specific inhibitor of DPP-4. Studies in experimental animals demonstrated that sitagliptin inhibits DPP-4 *in vivo* in this setting. Now the issues were: how does sitagliptin engage DPP-4 in people? What are the consequences of that engagement?

Target engagement is not a classic part of phase 1 clinical studies. The Merck approach was unusual and innovative; performing these studies at this early stage is reflective of a more modern clinical development program.

A phase 1 randomized, placebo-controlled, double-blind study enrolled 70 healthy male volunteers who received multiple oral doses of sitagliptin varying from 25 to 400 mg or placebo once a day for 10 days. In addition to the pharmacokinetic values discussed earlier, this study revealed the following. First, sitagliptin inhibited DPP-4 in blood. The degree of DPP-4 inhibition increased with increasing sitagliptin dose. At doses of sitagliptin equal to or greater than 100 mg once a day, inhibition of DPP-4 over 24 hours was at least 80 percent. Following a standard meal, concentrations of active GLP-1 were increased in the sitagliptin groups by about twofold compared with the placebo group. Sitagliptin was generally well tolerated and no evidence of hypoglycemia was observed. The results of this study established that sitagliptin worked in people as expected: DPP-4 was inhibited, inhibition increased with sitagliptin dose, and inhibition of DPP-4 resulted in increased active GLP-1 concentrations. No effect was seen on blood glucose levels. The final point is particularly important. Unlike insulin and sulfonylureas, sitagliptin does not cause hypoglycemia in healthy people or in patients with diabetes because it does not lower blood glucose levels when they are normal. Recall that sitagliptin elicits *glucose-dependent* insulin secretion via the action of active GLP-1. These findings fully justified moving forward.

THE ENVIRONMENT WAS COMPETITIVE: PHASE IB STUDIES

Merck was not alone is pursuing clinical studies of a gliptin—not alone and not ahead either. In reality, Novartis was about 2.5 years ahead of Merck and was pushing their drug candidate, vildagliptin, forward. In these drug development races, in which the

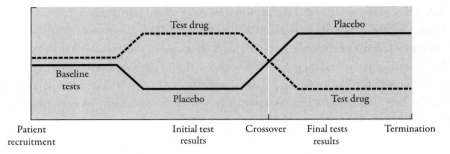

FIGURE 12.13 Clinical Crossover Study Design. In this type of study design, two randomized groups of patients are created. Initially, one group gets the test drug and the other the placebo. At the crossover point, the roles are exchanged; the test drug group gets placebo and the placebo group gets the test drug. The utility of the crossover design is that any meaningful differences in the nature of the two groups is nullified by the design. At the same time, there is a risk in this design. It is possible, for example, that there may be some carryover effect from the first period, before the crossover, into the second one, subsequent to the crossover.

complete biological profiles of competing product candidates are not fully defined, there is every reason to try to be the first to market (as any marketing expert will testify). Merck was faced with a daunting challenge of trying to narrow the time gap and, if possible, beat Novartis to the market.

So Merck physicians were faced with the task of devising a clinical strategy that would be both quick (efficient) and effective (generate an approvable NDA in the shortest time reasonably possible). Working through the usual sequence of phase 2A, phase 2B, and phase 3 studies was not going to suffice. A novel strategy was required—a streamlined, focused clinical program.

Merck choose a phase 1B experimental medicine strategy in which clinical trials that would ordinarily have been postponed until phase 2 were pulled forward to accelerate the clinical development process. Two studies were conducted. The first study was a randomized, double-blind, placebo-controlled, single-dose crossover study (Figure 12.13) that enrolled 58 *patients* with type 2 diabetes. Two doses of sitagliptin were studied: 25 mg and 200 mg (each dose strength was administered as a single dose). This was a study in which sitagliptin dose was correlated with a clinical end point: blood glucose control as measured by an oral glucose tolerance test. The goal here was to get a clinically relevant measurement early. Sitagliptin was shown to inhibit blood DPP-4 activity, increase blood active GLP-1, increase blood insulin, decrease blood glucagon, and improve glucose tolerance. Effects at the 200-mg dose were greater than those at the 25-mg dose. These results were exactly as expected. The near-maximal glucose-lowering effect of sitagliptin was achieved at doses that inhibited DPP-4 activity 80 percent or greater. Sitagliptin was generally well-tolerated in this study.

The second study was a randomized, placebo-controlled, double-blind, 4-week *safety study* in obese patients (obese patients are far easier to recruit than those with type 2 diabetes, and many patients with type 2 diabetes are obese or overweight). A high dose of sitagliptin was used in these studies—400 mg given as 200 mg twice a day—in an effort to demonstrate a substantial margin of safety for sitagliptin. Sitagliptin was generally well tolerated at this high dose. The outcomes of these two studies permitted Merck to go directly to phase 2B.

PHASE 2B STUDIES ESTABLISHED THE OPTIMAL DOSING REGIMEN

The key objectives of the phase 2B studies were to establish the dose and the dose regimen, once a day or twice a day, that optimizes blood glucose control in patients with type 2 diabetes. This dose and regimen would then be used in the large, critical phase 3 studies.

Two randomized, double-blind, placebo-controlled phase 2B studies were conducted. Patients with type 2 diabetes were dosed for 12 weeks in each case. The measure of blood glucose control was the level of glycated hemoglobin, HbA1c, a measure of long-term blood glucose control. In the first study, several doses of sitagliptin, ranging from 25 to 100 mg *once a day*, were studied as well as a 50-mg dose given *twice a day*. This approach allowed a direct comparison of 100 mg per day given once a day and in divided doses twice a day in the same study. In the second study, several doses of sitagliptin, ranging from 12.5 to 50 mg *twice a day*, were studied. The result was clear and favorable: 100 mg once a day gave optimal blood glucose control in these patients with type 2 diabetes. This result provided the necessary guidance for phase 3 studies. Note that these phase 2B studies were completed, from first patient enrolled through to study end and preliminary report generation, in 12 months, which is *fast*. A typical phase 2B trial could easily take twice as long.

HERE ARE THE PHASE 3 TRIALS THAT LED TO THE FIRST NDA

Five phase 3 clinical trials were included in the initial NDA filed with the FDA. It is worthwhile to have a look at these one at a time. I begin with two monotherapy clinical trials that were conducted simultaneously.

The first monotherapy trial was a randomized, placebo-controlled, double-blind study in which 741 patients with type 2 diabetes were treated with 100 mg or 200 mg of sitagliptin once a day for 24 weeks. The clinical end point was the level of glycated hemoglobin, HbA1c, as a measure of blood glucose control. This study was designed to mimic the anticipated way that sitagliptin would be used when on the market. After 24 weeks, HbA1c was reduced significantly at both sitagliptin doses compared with placebo. Insulin secretion and markers of pancreatic beta cell function were improved by sitagliptin. The drug was generally well tolerated.

The second monotherapy trial was similar in design to the first, with the exception that the duration of treatment was 18 weeks. A total of 521 patients were enrolled. Key clinical end points were the level of blood HbA1c and blood glucose level after a standardized meal. Results were consistent with those of the first trial. These two phase 3 clinical trials were critical for the sitagliptin program; monotherapy was anticipated to be an important way that sitagliptin would be used in medical practice. It was key to get the monotherapy indication in the drug label.

Note that phase 3 studies are designed to support specific claims in the product labeling. The nature of these claims is critical to the marketing organization as well as the research and development organization. The commercial side of the pharmaceutical company must market the product on the basis of the approved claims. Cooperation between the science side and the marketing side is highly important. Creating an approved drug is difficult, and one wants to see it used to the benefit of appropriate patients.

The third phase 3 clinical trial was also critical to approval. This randomized, placebo-controlled, double-blind trial evaluated the safety and efficacy of 100 mg sitagliptin once a day added to ongoing metformin therapy in 701 patients with type 2 diabetes controlled inadequately by metformin alone. As noted earlier, metformin is the most widely used oral drug to lower glucose in patients with type 2 diabetes. There was no doubt that addition of sitagliptin to patients on metformin would be a common use of the drug for those patients on metformin therapy but whose diabetes was controlled inadequately by it. It was critical to demonstrate that doing so would improve blood glucose control safely in this setting as measured by the level of blood HbA1c. The key result was that adding sitagliptin to ongoing metformin therapy reduced blood HbA1c, reduced fasting blood glucose levels, and reduced postmeal blood glucose levels. Here, too, sitagliptin was generally well tolerated by patients.

This clinical trial is of particular importance. It was critical for the approval of sitagliptin but also provided the *clinical* data required for the initial approval of the combination of sitagliptin and metformin in a single tablet. I return to this issue later.

The fourth phase 3 clinical trial in the initial NDA was nice to have but not critical to approval. It was a randomized, placebo-controlled, double-blind 24-week trial in which 353 patients were treated with 100 mg sitagliptin once a day as add-on therapy to pioglitazone. Here, too, the clinical trial was designed to explore the consequences of the way in which sitagliptin might be used after approval. As noted earlier, the glitazones have fallen out of favor, for safety reasons, in treatment of patients with type 2 diabetes, but this was not the case at the time of this trial. The results of this trial were consistent with those described for the metformin add-on trial.

The final phase 3 trial was a 12-week randomized, placebo-controlled, double-blind study of sitagliptin in 91 patients with type 2 diabetes with moderate or severe kidney function impairment (renal insufficiency), including patients with end-stage renal disease on dialysis. Because sitagliptin is eliminated through the kidneys, patients with moderate renal insufficiency were treated with 50 mg sitagliptin once a day; those with severe or end-stage renal insufficiency were treated with 25 mg once a day. These doses give blood concentrations of sitagliptin comparable with those achieved in patients with normal kidney function receiving 100 mg once a day. In this study, which was extended to 54 weeks, sitagliptin was generally well tolerated and provided effective glycemic control in patients with type 2 diabetes with both moderate and severe renal insufficiency, including patients on dialysis. In subsequent 54-week clinical trials, the efficacy and safety of sitagliptin in 426 patients with type 2 diabetes and renal insufficiency but not on dialysis was examined as well as 129 patients with type 2 diabetes and end-stage renal insufficiency on dialysis. The results of these studies were consistent with those just described.

Data from these five phase 3 trials were included together with earlier clinical trials, all preclinical data, and all information required about the sitagliptin molecule itself—stability, purity, manufacture—into an NDA that was submitted to the FDA and approved in October 2006. Merck had won the race: sitagliptin was first to market in the gliptin category. Januvia was born. This was the first hallelujah moment for sitagliptin. The approved indication for use is simple and compelling: Januvia is indicated as an adjunct to diet and exercise to improve glycemic control in adults with type 2 diabetes mellitus.

ADDITIONAL CLINICAL TRIALS SUPPORTED SAFE AND EFFECTIVE USE IN A VARIETY OF CLINICAL SETTINGS

Clinical studies of sitagliptin did not cease after the drug was approved for marketing. In fact, many clinical trials of sitagliptin continue to this day. The goal of further clinical studies was to gain additional information about the effects of sitagliptin in different treatment scenarios and to secure additional uses for the drug for which it may be marketed.

The studies that supported these goals were randomized, placebo-controlled, double-blind studies using sitagliptin at 100 mg once a day for a duration of 24 to 54 weeks. The level of blood HbA1c was the key clinical end point, although others were measured. These studies were constructed to compare sitagliptin with placebo against a family of background therapies, including metformin, sulfonylurea, sulfonylurea plus metformin, pioglitazone plus metformin, rosiglitazone plus metformin, insulin, and insulin plus metformin. The goal was to mimic additional likely scenarios in which sitagliptin would be used in the treatment of patients with type 2 diabetes. Sitagliptin proved safe and effective in each of these settings, justifying widespread use of the drug.

INITIAL COMBINATION THERAPY OF SITAGLIPTIN AND METFORMIN WAS EXPLORED FOR EFFICACY AND SAFETY IN CLINICAL TRIALS

Patients with diabetes frequently take a lot of drugs to control their blood glucose, to control their blood cholesterol (many patients with type 2 diabetes are on a statin), to treat the sequelae of diabetes, to treat the high blood pressure that many have, and so on. Compliance with dosing regimens is better when patients have to take fewer pills. As noted earlier, many patients with type 2 diabetes take both metformin and sitagliptin; sitagliptin as add-on therapy to metformin proved safe and effective. It was tempting to combine the two in a fixed ratio in a single tablet. Because metformin is not given to patients with renal insufficiency, the dose of sitagliptin in a combined tablet was clear at the outset: 100 mg per day. Metformin is usually given as 1,000 mg or 2,000 mg per day, administered in divided doses usually twice daily.

Tablets containing 50 mg sitagliptin and 500 mg metformin or 50 mg sitagliptin and 1,000 mg metformin were shown to be bioequivalent to the two drugs given separately. On that basis and the clinical data cited earlier, the combination was approved for treatment of patients with type 2 diabetes in April 2007. The combination is given twice a day and is marketed as Janumet. The initial indication for Janumet was for patients with type 2 diabetes in whom use of both agents is appropriate. This was the second hallelujah moment for sitagliptin.

It was also tempting to consider use of the sitagliptin/metformin combination (Janumet) as *initial therapy* rather than in an add-on scheme for patients with more severe hyperglycemia. The combination was explored in a 24-week, randomized, double-blind, placebo controlled study of 1,091 patients with type 2 diabetes. Six daily drug regimens were explored: 100 mg sitagliptin plus 1,000 mg metformin, 100 mg sitagliptin plus 2,000 mg metformin, 1,000 mg metformin alone, 2,000 mg metformin alone, 100 mg sitagliptin alone, and placebo. Medication was given in divided doses twice a day (with the exception of the sitagliptin-alone arm, in which sitagliptin was given 100 mg once a day with blinding maintained through a second "divided" placebo dose). This trial design provided for two comparisons of the combination against metformin alone, two comparisons of the combination against sitagliptin alone, and comparison of all regimens against placebo. The result was clear: the initial combination of sitagliptin and metformin provided substantial and additive improvement in glycemic control compared with either metformin alone or sitagliptin alone and was generally well tolerated in these patients. These results justify the use of Janumet as initial therapy in appropriate patients with type 2 diabetes.

A subsequent study of initial combined sitagliptin-and-metformin therapy included 1,250 patients in a randomized, double-blind clinical trial in which sitagliptin was given with various doses of metformin administered in divided doses twice a day for 18 weeks. The outcome was impressive: sitagliptin plus metformin proved more effective than metformin alone in reducing the level of HbA1c (−2.4 percent vs. −1.8 percent), and more patients treated with the combination had an HbA1c level less than 7 percent, which was the goal. In addition, reductions in fasting blood glucose levels were greater with the combination than with metformin alone (−3.8 mmol/L vs. −3.0 mmol/L). These results were confirmed and amplified in subsequent clinical trials.

Clinical trials continued and continue, including studies in special populations (age, ethnicity) and studies in patients with prediabetes. Of particular note is an ongoing cardiovascular outcomes study involving more than 14,000 patients that was initiated in 2008. The first meaningful data are expected in 2015. The cost of this study is likely to total several hundreds of millions of dollars.

Last, it is possible to pool data from multiple studies of sitagliptin to get a global picture of the safety and tolerability of it. The latest effort to do so collected data from 19 placebo (or active drug)-controlled, double-blind studies involving 10,246 patients with type 2 diabetes. Here is the comforting conclusion:

> Summary measures of overall adverse events were similar in the sitagliptin and non-exposed [to sitagliptin] groups, except for an increased incidence of drug-related adverse events in the non-exposed group. Incidence rates of specific adverse events were also generally similar between the two groups, except for increased incidence rates of hypoglycemia, related to the greater use of sulfonylurea, and diarrhea, related to the greater use of metformin, in the non-exposed group, and constipation in the sitagliptin group. Treatment with sitagliptin was not associated with an increased risk of major adverse cardiovascular events.[13]

Sitagliptin is a remarkably safe drug.

Fittingly, in 2011, Ann E. Weber and Nancy A. Thornberry were awarded the Discoverers Award, the highest honor from the Pharmaceutical Research and Manufacturers of America (Figure 12.14). This is the only occasion in the 24-year history of this prize that women alone have been the winners.

Nancy Thornberry, Ann Weber, Keith Kaufman, Peter Stein, and Gary Herman were recipients of the Merck Director's Scientific Award for sitagliptin.

Sitagliptin Was First But Is Not Alone

Merck was hardly the only player in the game of discovering inhibitors of DPP-4, but it was the first to market. Subsequently, a modest flow of gliptins, all inhibitors of DPP-4, has been approved in the United States and elsewhere. They include vildagliptin (not available in the United States but marketed elsewhere as Galvus, Zomelis), saxagliptin (Onglyza), and linagliptin (Tradjenta). Alogliptin (Nesina) is approved in the United States and Japan. Each of these will find some utility in the control of blood glucose levels in type 2 diabetes and in reducing the risk of the microvascular complications of that disease.

The gliptins are an important addition to the family of tools available to prescribing physicians to improve the health and well-being of patients with type 2 diabetes throughout the world. We are better off for having them.

There Is a Search for Second-Generation Gliptins with Properties Superior to Those of Sitagliptin

Under the leadership of Thornberry and Weber, work at Merck continues to optimize the efficacy and safety of the gliptins for the treatment of patients with type 2 diabetes. One issue is adherence to prescribed dosage regimens. Medication adherence in chronic diseases—hypertension, type 2 diabetes, arthritis, and others—may average only about 50 percent. No matter how good the drug is, it cannot be optimally effective if patients do not take it as prescribed. In general, the simpler the dosage regimen and the fewer pills prescribed, the better patient adherence. Once-weekly dosing with a gliptin may have an advantage over sitagliptin for patients with type 2 diabetes.

The detailed structure of DPP-4 was not available to Merck scientists at the time of discovery of sitagliptin (see Figure 12.11). It has played an important role for the design of subsequent DPP-4 inhibitors, including a novel DPP-4 inhibitor known as *MK-3102*, which is an amazing molecule. MK-3102 is roughly 10 times as potent as sitagliptin, has a very long blood half-life suitable for once-weekly dosing at a dose one-quarter that of sitagliptin, and is largely free of metabolism. It is excreted by the kidney but no dose reduction for patients with mild or moderate renal insufficiency is anticipated. It is not certain that MK-3102 will reach the marketplace, but it is difficult not to be optimistic. MK-3102 is in phase 3 clinical trials. There may yet be another hallelujah moment down the road for Merck in the treatment for patients with type 2 diabetes!

Notes

CHAPTER 1 SEDUCED BY DRUG DISCOVERY

1. At the time, the research facilities of Merck and Company were known as the Merck, Sharp, and Dohme Research Laboratories. The name was subsequently simplified to Merck Research Laboratories and I have elected to use the current term. I frequently refer to Merck Research Laboratories simply as Merck Research or Merck.

2. Ribonuclease A is a protein comprised of a single chain of 124 amino acids. Ribonuclease A is cleaved by the enzyme subtilisin into ribonuclease S, which is 104 amino acids long; and S-peptide, which comprises the other 20 amino acids. Hirschmann and Denkewalter used solution chemistry in their synthesis of ribonuclease S whereas Merrifield used solid-state chemistry, which he invented, in his synthesis of ribonuclease A.

3. After his retirement from Merck at age 65, Ralph Hirschmann enjoyed a fine postindustry career as a professor of chemistry at the University of Pennsylvania. Ralph passed away in mid 2009 from kidney failure at age 87.

4. P. Roy Vagelos and Louis Galambos, *Medicine, Science, and Merck* (New York: Cambridge University Press, 2004). See chapters 5 through 7, in particular.

5. "The Story of Januvia," www.innovation.org.

CHAPTER 2 THE SMALL MOLECULES OF LIFE

1. Primaxin was a breakthrough antibiotic the discovery and development of which is described in chapter 9. Fludalanine is a failed antibiotic candidate described in chapter 11.

2. Molecules that transmit messages from one neuron to another or from a neuron to a muscle cell are called *neurotransmitters*. For example, acetylcholine and serotonin are neurotransmitters

in both the sea snail *Aplysia* and humans. Humans have a greater diversity of neurotransmitters than sea snails, however.

3. This number is known as Avogadro's number, after the Italian chemist Amadeo Avogadro, who proposed in 1811 that, at constant temperature and pressure, the number of atoms or molecules of a gas is proportional to the volume of the gas and is independent of the nature of the gas. French chemist Jean Perrin first determined a reliable value for Avogadro's number. Perrin was awarded the Nobel Prize in Chemistry in 1926 largely for this work. The value of Avogadro's number has since been refined a number of times.

4. Lawrence J. Henderson, *The Fitness of the Environment* (New York: Macmillan, 1913).

5. Nick Lane, *Oxygen: The Molecule That Made the World* (New York: Oxford University Press, 2002).

6. Molecular oxygen serves other purposes, as well. In many of the chemical transformations in metabolism, one or two atoms of molecular oxygen are incorporated directly into molecules. For example, the amino acid tyrosine contains an atom of oxygen that is not present in its precursor amino acid phenylalanine. That atom of oxygen comes from O_2.

CHAPTER 3 PROTEINS: MOLECULAR WONDERS IN THREE DIMENSIONS

1. The structures in Figures 3.5 through 3.8 are derived from a high-resolution X-ray structure of human renin by A. R. Sielecki, K. Hayakawa, M. Fujinaga, M. E. Murphy, M. Fraser, A. K. Muir, C. T. Carilli, J. A. Lewicki, J. D. Baxter, and M. N. James, "Structure of Recombinant Human Renin, a Target for Cardiovascular-Active Drugs, at 2.5 A Resolution," *Science* 243 (1989): 1346–1351.

2. These compositions are taken from Roald Hoffmann, *The Same and Not the Same* (New York: Columbia University Press, 1995).

CHAPTER 4 PROTEINS PERFORM MULTIPLE FUNCTIONS: ENZYMES, RECEPTORS, ION CHANNELS

1. Note that in the case of the digestion of proteins in the gut, proteins are both catalysts, enzymes, and substrates—dietary proteins. The enzymes are known generically as *proteases*, enzymes that digest proteins. One might ask: why do the proteases not digest themselves because they are proteins? The answer is that the proteases are quite specific for the amino acid sequence on which they act. Each protease lacks (or hides) that amino acid sequence so it does not act on itself.

CHAPTER 5 DRUG DISCOVERY AND DEVELOPMENT: THE ROAD FROM AN IDEA TO PROMOTING HUMAN HEALTH

1. All drugs used in human medicine have at least two names. For example, captopril is a nonproprietary name suggested by Squibb subject to approval by the U.S. Adopted Names Council. The captopril molecule is known by that name worldwide. Captopril is marketed as Capoten, a trade name that Squibb has trademarked. The FDA must approve trade names in the United States. Some drugs are marketed under many names. Enalapril is marketed as Vasotec in the United States and as Renitec in Europe. In extreme cases, one drug may be marketed under 10 or more trade names around the world.

2. Mutations are, in one way or another, changes in DNA structure in the genome. Because many genes of DNA code for the primary structure of proteins, mutations may affect protein structure and, hence, protein biological function. We encountered one case earlier, that of sickle cell anemia (see chapter 3).

3. A genome is the totality of the genetic information, usually carried as DNA, in a living organism. To define a genome, one must determine the sequence of nucleotide base pairs of the genetic DNA. In humans, this DNA is carried on 23 pairs of chromosomes of varying length. The total number of base pairs is about 3 billion. The first draft of the human genome was published in 2001. It has subsequently been refined, and the technology for determining genome structure has improved remarkably since then.

4. Here is how one gene can code for more than one protein. Some genes are composed of alternating protein coding and noncoding regions, termed *exons* and *introns*, respectively. The exons contain the information required to specify amino acid sequence in proteins; the introns do not. During the process of protein synthesis, the introns are spliced out, bringing the exons together. Consider a simple case in which the gene contains one intron: exon1–intron–exon2. In principle, this gene can code for two proteins, one coded for by exon1 and one coded for by exon1–exon2.

5. Typically, successful molecules in the current competitive world have measures of potency in the nanomolar range. For example, values of the thermodynamic inhibition constant (K_i) or the concentration required to achieve 50% inhibition (IC_{50}) will be 1×10^{-9} M or less.

6. As always, there are exceptions. Some successful drugs for treatment of cancer are "dirty"—that is, they hit more than one molecular target. Because, in general, cancer is the consequence of multiple somatic mutations and not just one, this makes a certain amount of sense.

7. Good Laboratory Practices (GLP) include a set of very detailed regulations that must be followed during the course of preclinical development. FDA personnel audit adherence to GLP standards during the course of a visit to the pharmaceutical company's research and development site or sites. Failure to adhere results in an action that can affect adversely the ability to gain approval for the drug. In addition to GLP, there are Good Clinical Practices and Good Manufacturing Practices standards. The pharmaceutical industry is very heavily regulated, for good reason.

CHAPTER 6 FINASTERIDE: THE GARY AND JERRY SHOW

1. The key reference to the original literature is R. E. Peterson, J. Imperato-McGinley, T. Gautier, and E. Sturla, "Male Pseudohermaphroditism Due to Steroid 5 α-Reductase Deficiency," *American Journal of Medicine* 62 (1977): 170–191. The photographs that appear as Figures 6.3 through 6.5 are taken from this publication. Note that male pseudohermaphroditism is not limited to this Dominican Republic community. It was first recognized and followed up scientifically in this group.

2. J. Eugenides, *Middlesex: A Novel* (New York: Farrar, Straus and Giroux, 2002).

3. Affected individuals are not entirely anatomically normal after puberty. They have a blind vaginal canal and the urethra exits at the base of the penis, a condition known as *hypospadias*.

4. The prostate size was measured by acquiring access surgically to the prostate in anesthetized dogs. The prostate dimensions were measured with calipers, and the volume was calculated. There is a more elegant and easier way to do this: transrectal magnetic resonance imaging.

It avoids the surgery, the prostate is visualized directly, and the measurement can be repeated over time to follow the rate of prostate size reduction. Our magnetic resonance imaging expert, Sheila Cohen, subsequently carried out a study of finasteride action in dogs using this technique. The results were excellent and fully in accord with those acquired more traditionally.

5. For a summary of the medicinal chemistry effort, see G. H. Rasmusson, G. F. Reynolds, N. G. Steinberg, E. Walton, G. F. Patel, T. Liang, M. A. Casieri, A. N. Cheung, J. R. Brooks, and C. Berman, "Azasteroids: Structure–Activity Relationships for Inhibition of 5-Alpha Reductase and of Androgen Receptor Binding," *Journal of Medicinal Chemistry* 29 (1986): 2298–2315.

6. For a summary of the clinical program, see E. Stoner, "The Clinical Development of a 5-Alpha Reductase Inhibitor: Finasteride," *Journal of Steroid Biochemistry and Molecular Biology* 37 (1990): 375–378.

CHAPTER 7 BASIC RESEARCH, SNAKE VENOMS, AND ACE INHIBITORS: ONDETTI, CUSHMAN, AND PATCHETT

1. The actual ligands to the zinc atom (Zn^{2+}) in ACE and carboxypeptidase A are derived from the -SH and -COOH groups by loss of a proton. Both the -SH and the -COOH groups are weak acids and can lose a proton under physiological conditions to form the corresponding anions, $-S^-$ and $-CO_2^-$. These anions bind to the positively charged zinc atom much more tightly than the uncharged structures do.

CHAPTER 8 STATINS: PROTECTION AGAINST HEART ATTACKS AND STROKES—HALLELUJAH

1. For a detailed summary of the statins, see J. J. Li, *Triumph of the Heart: The Story of Statins* (New York: Oxford University Press, 2009).

2. Baychol was withdrawn from the market in 2001 after 52 reported deaths related primarily to rhabdomyolysis.

3. In fact, Konrad Bloch, about whom much more follows later, has argued cogently that cholesterol has been optimized for its functions through a process of molecular evolution. See Konrad Bloch, *Blondes in Venetian Paintings, the Nine-Banded Armadillo, and Other Essays in Biochemistry* (New Haven, CT: Yale University Press, 1994), 14–36.

4. Feodor Lynen made a number of important discoveries. He was famous for his wit, his determination, and his prodigious capacity for alcohol. I met Lynen at a small scientific meeting—a Gordon Conference—in rural New Hampshire many years ago. During a free afternoon, he and a few others had taken a hike and gotten lost. Lynen made his way to a farmhouse, announced who he was, and asked for a beer, which was forthcoming. A rescue party returned him to the meeting (and liberated the farmhouse), and my wife and I joined him for a predinner drink. The drinking lasted through the dinner hour, with Lynen doing most of the talking and more than his share of the drinking. The evening scientific talks were a bit hazy in my mind. After five days of this, Lynen gave a splendid talk on the final day of the meeting.

5. The quote is from Daniel Steinberg, *The Cholesterol Wars: The Skeptics vs. the Preponderance of Evidence* (New York: Academic Press, 2007), 1.

6. This turns out to be quite important for the history of statins. By coincidence, scientists at Beecham Pharmaceuticals in England discovered mevastatin in a related mold at about the

same time that Endo did in Japan. They demonstrated that substantial inhibition of cholesterol biosynthesis in rats by mevastatin did not result in a lowering of blood LDL. They concluded, in a stunning error, that if mevastatin did not work in rats, it would not work in other species either. They did not contest this conclusion through work with mevastatin in other species and closed the project. Use of an inappropriate animal model excluded Beecham from participation in the statin field.

7. A. Yamamoto, H. Sudo, and A. Endo, "Therapeutic Effects of ML-236B in Primary Hypercholesterolemia," *Atherosclerosis*, 35 (1980): 259–266.

8. For revealing stories of the development of lovastatin and simvastatin, see Steinberg, *The Cholesterol Wars*, chapter 8; and P. Roy Vagelos and Louis Galambos, *Medicine, Science, and Merck* (New York: Cambridge University Press, 2004).

9. Carl Hoffman was a member of Karl Folkers group that discovered mevalonate back in 1956 (see page 123). Thus, his entire career at Merck evolved around that molecule. Carl passed away a few days after the approval of lovastatin.

10. Vagelos and Galambos, 135.

11. A. B. Goldfine, "Is It Really Time to Reassess Benefits and Risks?" *New England Journal of Medicine* 366 (2012): 1753–1755 (quote appears on p. 1755).

CHAPTER 9 THE PERILS OF PRIMAXIN

1. The term *antibiotic* is general and includes antibacterials, antivirals, and antiparasitics. However, drugs active against bacterial infections are quite often referred to simply as antibiotics, and I am going to use this term throughout this chapter.

2. H. Kropp, J. S. Kahan, F. M. Kahan, J. Sundelof, G. Darland, and J. Birnbaum, *Abstracts XVI, Interscience Conference on Antimicrobial Agents and Chemotherapy*, no. 299 (Chicago, IL: 1976).

3. J. S. Kahan, F. M. Kahan. R. Goegelman, S. A. Currie, M. Jackson, E. O. Stapley, T. W. Miller, A. K. Miller, D. Hendlin, S. Mochales, S. Hernandez, H. B. Woodruff, and J. Birnbaum. "Thienamycin, a New Beta-Lactam Antibiotic: I. Discovery, Taxonomy, Isolation, and Physical Properties," *Journal of Antibiotics (Tokyo)* XXXII (1979): 1–12.

4. The β-lactam ring of thienamycin is unusually susceptible to nucleophilic attack, including that by buffers used to maintain solution neutrality. Tertiary amine buffers were chosen that were hindered sterically from attacking the β-lactam. As discussed later in the chapter, thienamycin has a primary amino group on a side chain that acts as a nucleophile toward the β-lactam ring on another thienamycin molecule, causing the concentration-dependent instability.

5. Most of the structural work was done on the *N*-acetylmethyl ester of thienamycin, a stable molecule.

6. G. Albers-Schönberg, B. H. Arison, O. D. Hensens, J. Hirshfield, K. Hoogsteen, E. A. Kaczka, R. E. Rhodes, J. S. Kahan, F. M. Kahan, R. W. Ratcliffe, E. Walton, L. J. Ruswinkle, R. B. Morin, and B. G. Christensen, "The Structure and Absolute Configuration of Thienamycin," *Journal of the American Chemical Society* 100 (1978): 6491–6499.

7. The structure of the thienamycin molecule provided the basis for the name: The sulfur atom outside the ring system provides the "thi," the double bond in the five-member ring provided the "ene," and the lactam structure provided the "am." Very neat.

8. H. Kropp, J. G. Sundelof, J. S. Kahan, F. M. Kahan, and J. Birnbaum. MK0787 (N-formimidoyl thienamycin): Evaluation of in Vitro and in Vivo Activities," *Antimicrobial Agents and Chemotherapy* 17 (1980): 993–1000.

9. Ibid.

10. H. Kropp, J. G. Sundelof, R. Hadju, and F. M. Kahan, "Metabolism of Thienamycin and Related Carbapenems by the Renal Dipeptidase, Dehydropeptidase," *Antimicrobial Agents and Chemotherapy* 22 (1982): 62–70.

11. Ibid.

CHAPTER 10 AVERMECTINS: MOLECULES OF LIFE BATTLE PARASITES

1. The quotation comes from W. C. Campbell, "The Genesis of the Antiparasitic Drug Ivermectin," in *Inventive Minds*, ed. R. J. Weber and D. N. Perkins (Oxford University Press, 1992), 194–217.

2. My personal involvement in the avermectin story was entirely peripheral. Ching C. Wang did important early studies on the mechanism of action of the avermectins and was part of my group at Merck. For the most part, I was an observer. The substance of this chapter comes from the writings of William C. Campbell and personal correspondence with him. With his unfailing help, I believe that I have gotten the story about right. For a current and careful history of ivermectin development, see William C. Campbell, "History of Avermectin and Ivermectin, with Notes on the History of Other Macrocyclic Lactone Antiparasitic Agents," *Current Pharmaceutical Technology* 13 (2012): 853–865.

3. Although Max Tishler had left Merck for Wesleyan University before my arrival at Merck, I did get to know him. It was illuminating. If Max had a question, he would phone me in the morning with it. If I did not have an answer by mid afternoon, I would get a call from an annoyed Max wondering why not. I learned to put Max's issues at the top of my priority list and made an effort to get back to him before he got back to me. He must have been an interesting leader for whom to work.

4. I am indebted to Jerry Birnbaum for the sequence of events that led to Kitasato cultures arriving in Rahway, New Jersey.

5. To be precise, the definition of ivermectin is a mixture containing at least 80 percent 22, 23-dihydro-avermectin B1a and not more than 20 percent 22, 23-dihydro-avermectin B1b.

6. This needs to be refined a bit. There is no evidence of a Collie being affected adversely when given the recommended dose of ivermectin for heartworm prevention. There is a safety problem when Collies are given a high ivermectin dose, yet one that is tolerated by other dog breeds. Therefore, tolerance to ivermectin in some Collies is reduced compared with that of other species.

7. To be precise, the definition of abamectin is a mixture containing at least 80 percent of avermectin B1a and not more than 20 percent of avermectin B1b.

8. The story of the clinical development of ivermectin is taken largely from a report compiled by Dr. William C. Campbell.

9. M. A. Aziz, S. Diallo, I. M. Diop, M. Lariviere, and M. Porta, "Efficacy and Tolerability of Ivermectin in Human Onchocerciasis," *Lancet* 2(1982): 171–173.

10. Quotation from P. Roy Vagelos and L. Galambos, *Medicine, Science and Merck* (New York: Cambridge University Press, 2004), 250.

CHAPTER 11 FLUDALANINE: NICE TRY BUT NO HALLELUJAH

1. I was not a direct participant in the fludalanine work but was able to observe what was going on from some distance. I did not have the occasion to understand some aspects of the project, and the passage of time dimmed some of what I did understand. I am indebted to Fred Kahan, a major player in this tale, for setting me straight on a number of issues.

2. The escape of Kollonitsch from Hungary is quite extraordinary. I am indebted to Steve Marburg for the following story. At a time of upheaval in Hungary under the Communists, Kollonitsch judged that there was a window of opportunity to escape into Austria. He collected his wife, his children, and a suitcase full of cash and traveled to the Austrian border, where he was met by a submachine gun-carrying border guard. Kollonitsch handed the suitcase of cash to the guard, asserted that it contained 50,000 forints and asked the guard to count it to be sure while he and his family walked across the border. Partway across, the guard ran up to them and stated that it was all there! Kollonitsch and family walked on and eventually found their way to the United States, aided by a prominent American chemist from Purdue University, Herb Brown.

3. A maximum response in a dose–response profile generally reflects the saturation of some enzyme, receptor, or ion channel by the drug. For example, when a dose of an enzyme inhibitor is sufficient to complex with, say, 99 percent of the target enzyme, further increases in the dose will have a negligible effect. The enzyme is said to be saturated with inhibitor.

CHAPTER 12 DIABETES BREAKTHROUGHS: JANUVIA AND JANUMET

1. In your reading, you may encounter the terms *blood glucose, plasma glucose,* and *serum glucose.* They basically mean the same thing. Plasma is blood in which the red blood cells have been removed. Serum is derived from plasma by permitting it to clot and removing the clot. I am going to use the term *blood glucose* throughout.

2. Insulin action has other consequences for human physiology including appetite suppression and weight control.

3. The basic point here is to maintain blood glucose at a near-constant-level (homeostasis) by correcting excursions in either direction—too much through the action of insulin or too little through the action of glucagon. There is pathology associated with both excursions. In addition, glucose is the only metabolic fuel for the brain. In chemistry terms, the normal level of blood glucose is about 5 mM (millimolar).

4. The 1923, the Nobel Prize in Physiology or Medicine was awarded to Banting and John Macleod, a leading figure in diabetes in Canada. Macleod had made available the laboratory in which Banting and Best did their seminal work, and also participated in furthering their work in important ways. However, Banting was furious that Macleod had shared in the Nobel Prize rather than Best. This is one of the most controversial Nobel Prizes awarded; arguments continue.

5. Michael Bliss, *The Discovery of Insulin: 25th Anniversary Edition* (Chicago, IL: University of Chicago Press, 2007), 11.

6. The metabolic link between obesity and type 2 diabetes is partially understood. Free fatty acids, the breakdown products of fats, elicit insulin resistance. That is, insulin works poorly to influence the uptake of blood glucose in muscle and fat cells and to convert liver glucose into its storage form, glycogen.

7. There was a form of insulin developed that could be taken by inhalation. Surprisingly, patients with diabetes pretty much turned up their noses at this product and it was eventually withdrawn from the market as a result of lack of sales.

8. The glitazones work by activating a nuclear receptor known as peroxisome proliferator-activated receptor-type gamma (PPARγ). Once activated, it complexes with another nuclear receptor known as the retinoid X receptor (RXR) and the complex affects protein synthesis by linking to specific sites on DNA.

9. GLP-1 acts through its receptor on cell membrane surfaces. GLP-1 agonists act on the same receptor with the related physiological consequences.

10. Mice lacking DPP-4 are said to be *knockout mice*. The technologies of genetic engineering permit scientists to eliminate one or more genes from the genomes of many organisms. If you eliminate the gene, you also eliminate the protein product of that gene. The fact that mice lacking DPP-4 are healthy, strongly implies that an inhibitor of DPP-4 may be safe. But then again, mice are not people, and a worry remained.

11. Here are the structures of PBD-1 and PBD-2:

PBD-1   **PBD-2**

12. Nancy A. Thornberry and Ann E. Weber, "Discovery of JANUVIA (Sitagliptin), a Selective Dipeptidyl Peptidase IV Inhibitor for the Treatment of Type 2 Diabetes," *Current Topics in Medicinal Chemistry* 7 (2007): 557–568. This article provides an excellent summary of events in the discovery of sitagliptin.

13. D. Williams-Herman, S. S. Engel, E. Round, J. Johnson, G. T. Golm, H. Guo, B. J. Musser, M. J. Davies, K. D. Kaufman, and B. J. Goldstein, "Safety and Tolerability of Sitagliptin in Clinical Studies: A Pooled Analysis of Data from 10,246 Patients with Type 2 Diabetes," *BMC Endocrine Disorders* 10 (2010): 7.

Index